Mute Vol 2 #11

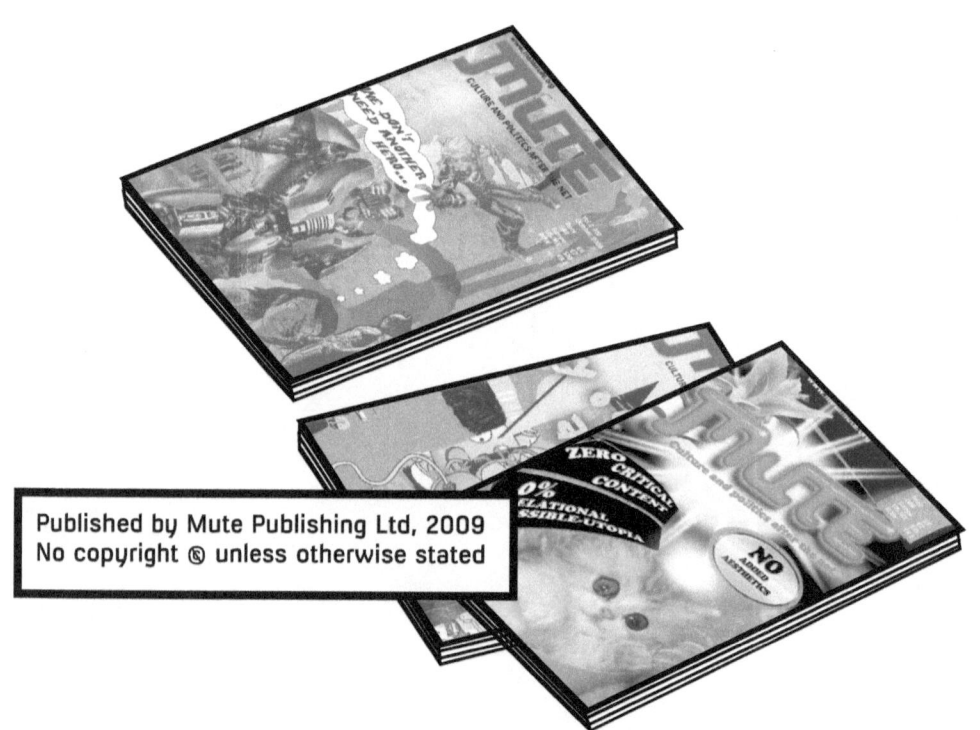

Published by Mute Publishing Ltd, 2009
No copyright ® unless otherwise stated

MUTE Vol 2 #11
SPRING ISSUE – MARCH '09

EDITOR
Josephine Berry Slater <josie@metamute.org>

EDITORIAL BOARD
Josephine Berry Slater, Matthew Hyland <infuriant@autistici.org>, Anthony Iles <anthony@metamute.org>, Demetra Kotouza <demetra@inventati.org>, Hari Kunzru <hari@metamute.org>, Melancholic Troglodytes <meltrogs1@hotmail.com>, Pauline van Mourik Broekman, Benedict Seymour <ben@metamute.org>, Stefan Szczelkun <szczels1@ukonline.co.uk> and Simon Worthington

MUTE PUBLISHING ADVISORY BOARD
Ceri Hand, Sally Jane Norman, Sukhdev Sandhu and Andy Wilson

PUBLISHERS
Pauline van Mourik Broekman <pauline@metamute.org>
Simon Worthington <simon@metamute.org>

ISSUE DESIGN
Laura Oldenbourg <laura@metamute.org>

ADVERTISING & MARKETING
Lois Olmstead <lois@metamute.org>
T: +44 (0)7791284039

WEBSITE
Metamute.org is powered by Drupal and CiviCRM FLOSS Software, with additional software services by our very own OpenMute http://openmute.org

TECH SUPPORT
Web infrastructure: Darron Broad
<darron@kewl.org>

PROJECT ASSISTANT CO-ORDINATOR
Caroline Heron <caroline@metamute.org>

INTERNS
Olga Panadés & Paul Graham

OFFICE
Mute, Unit 9, The Whitechapel Centre,
85 Myrdle Street,
London E1 1HQ, UK
T: +44 (0)20 7377 6949
F: +44 (0)20 7377 9520
email: <mute@metamute.org>

SUBSCRIPTIONS
Howard Slater
T: +44 (0)20 7377 6949
F: +44 (0)20 7377 9520
email: <subs@metamute.org>
web: http://www.metamute.org/subs/

DISTRIBUTION UK
Central Books,
99 Wallis Road,
London, E4 5LN
T: +44 (0)20 8986 4854
F: +44 (0)20 8533 5821

DISTRIBUTION NORTH AMERICA
Please contact:
Lois Olmstead <lois@metamute.org>
T: +44 (0)7791284039
or visit http://www.moreismore.net

CONTRIBUTING
Mute welcomes contributions of all kinds. Email <mute@metamute.org> with your ideas

You can also publish on Mute's website (http://www.metamute.org). Post news, texts, events and comments, or upload media to the Mute Public Library http://pl.metamute.org

The views expressed in Mute and Metamute are not necessarily those of the publishers or service providers

Mute is published in the UK by Mute Publishing Ltd. and printed by OpenMute (http://openmute.org) print on demand (POD) book services in the USA and UK

COVER
Theo Michael <mrtheomichael@yahoo.co.uk>

SPECIAL THANKS
To Matthew Hyland for writing the editorial – no mean feat this issue!

ISSN 1356-7748 - 211
ISBN 978-1-906496-26-5

Mute is supported by
Arts Council England

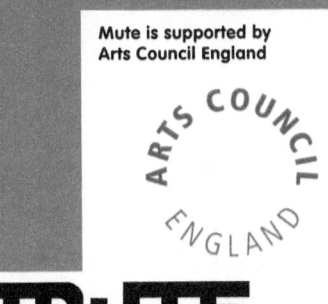

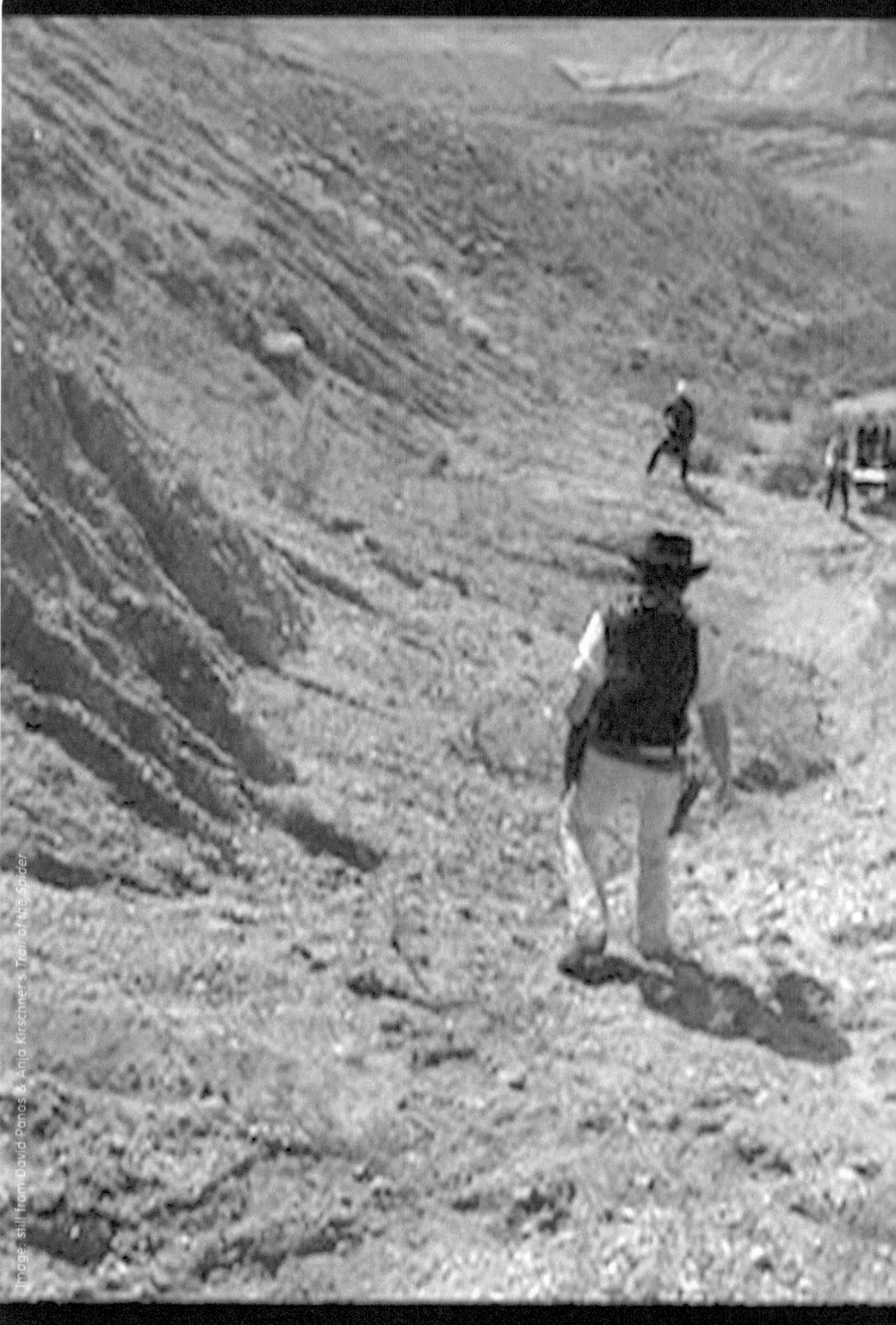

CONTENTS

6   **EDITORIAL: EMPATHY FOR THE DEVIL**
by Matthew Hyland

16   **BURDENED BY THE ABSENCE OF THE BILLIONS?**
Howard Slater reviews Frére Dupont's *Species Being and Other Stories* and considers the revolutionary repressed

28   **HE'S NOT BEYOND GOOD AND EVIL**
Nina Power asks if there's a point to Paolo Virno's unhappy human

36   **MONSTROUS PLANS & GOOD HABITATS**
Mark Crinson queries suggestions that anti-colonisation struggles were also about architecture

46   **THE POLITICAL IMMUNITY OF DISCOURSE**
Erik Empson reads Roberto Esposito and questions the revolutionary potential of philosophy

60   **WISHFUL THINKERS OF THE CALAMITY BAZAAR**
John Barker says the time to attack the fantasy world of capitalist spin-doctors is now

76   **THE WHO AND WHOM OF LIBERTY TAKING**
Peter Linebaugh asks how it's possible to discuss liberty in the absence of equality

86   **DUCK! YOU REGENERATION SUCKER**
Neil Gray watches David Panos and Anja Kirschner's *Trail of the Spider* and finds history repeating

96   **THE SLEEP OF REALISM PRODUCES MONSTERS**
Andrew Fisher questions documentary maker Adam Curtis' claims to 'realism' and political neutrality

EDITORIAL
EMPATHY FOR THE DEVIL

Lobotomist Walter Freeman to patient under local anaesthetic: *What's going through your mind right now?*
Patient: *A knife*.
– Quoted in Howard Dully's, *My Lobotomy: A Memoir*, 2007, (ghost-writer Charles Fleming)

How tired I am after another year of denunciation!
– Myles na gCopaleen

In this issue of *Mute*, Nina Power reviews the book in which Paolo Virno, that prince among humanist scholars, liquidates human history by declaring it finally to have achieved identity (or perhaps one should say 'parity', as between euro, dollar and pound?) with biological Human Nature. The sceptical reader might wonder why *Mute* has turned its high-strung attention towards a work with the word 'innovation' in the title: after all this is not Lucy Kellaway's *Financial Times* column on 'management twaddle'. But Virno, the 'dissociated' ex-*operaista* and exquisite professor of 'ethics of communication', has pulled off a feat quite beyond the powers of his anti-historicist *confrères* in the 'life sciences', one which forces us to reload the battered 'arms of criticism' whether we like it or not.

Certainly our peer-reviewed friends scorch from the earth every historical (which is to say material) conception of social and linguistic reality when they fill *Nature*, *Neuron* and *The New Scientist* with discoveries like: 'neural patterns determine how we [*sic*] judge and sentence criminals', or, 'genetic factors explain depression / musical talent / enjoyment of fast food', or, 'men are genetically programmed to prefer women with symmetrical faces' [nb. *NONE of these is made up. The editorialist will wade through the science comics and find the references on request at a reasonable hourly rate, plus expenses*]. And this kind of 'thinking' currently enjoys a blazing ascendancy in public policy and private research funding. But it has always awaited a Virno for its *Aufhebung*, the moment when it no longer stands opposed to stubborn notions of historical time, but can claim to be their *abolition and realisation*. This is what Virno delivers. As Nina Power paraphrases, he proclaims 'the contemporary

Matthew Hyland

multitude' to be 'perhaps the first truly historiconatural being'. Thus not only does history consign itself to the synchronous world of biology, the scope of the neurobehaviourists' explanatory power expands dramatically: from *cases* within the field of the social to social totality itself. Never mind that this sort of leap from clinically-contrived microphenomenon to general social soothsaying resembles nothing so much as the medieval medical theory of Analogy or Sympathetic Magic: its mechanistic directness actually entails the utmost superstitious abstraction. Perhaps this is why Virno, like the 'innovators' of Victorian positivism before him, ends up wrestling ignominiously with the 'Problem of Evil'.

Also featured in this slim (but definitely not *Changed4Life*) volume is Erik Empson on Roberto Esposito, whose book *Bios* is listed on Amazon under the category (look away now if you don't want to see THE END) *posthumanities* (as in 'posthumous', presumably). 'A sympathetic view of Esposito's book', writes Empson, 'would be that he aims to [...] produce an affirmative form of biopolitics based on a nuanced philosophical understanding.' One can only quiver in dread, then, at the thought of an *unsympathetic* view. *An affirmative biopolitics!* Has the Five-A-Day/Every Child Matters/Biometric Database/Schengen/Homeland Security meta-state rested for one moment in the last dozen years in the effort to bring this about (see *Mute*, Vol 2 #9)?!

Esposito seems to want to quarrel with the anti-Foucauldian biopolitics of Agamben's *Homo Sacer*, the only formulation other than that of the Sozialistisches Patientenkollektiv (SPK, http://www.spkpfh.de/) which is anywhere near *negative* enough to reflect lived misery. This debate is skewed a bit by Agamben's own 'messianic' affirmation of the absolutely negative conclusion he arrives at (extermination camp = *nomos* of the modern); more importantly, philosophy flinches from the obvious real world application: *capital* strives to reduce *labour* to 'bare life', i.e. labour power 'without qualities' and subject to 'sovereign decision' over its life or death, but cannot do so because it depends on qualified (as in 'with qualities', not 'clutching a diploma') human labour as source and measure of value.

Neither Agamben nor the SPK really grasps this basic contradiction, but at least both get as far as identifying biopolitics with what Midnight Notes writers have long called *thanatopolitics*, the politics of death. Yet Esposito seems intent on reversing even this limited progress. He throws out Agamben's account of fascism as generalisation of the 'sovereign exception' that constitutes political power: instead, fascism is a nebulous 'absolute normativization of life' (i.e. little more than 'authoritarianism', or a lot of pesky rules). In the same vein, *contra* Agamben but *pro* Foucault (and also, although Empson prefers to group him

Editorial

with Agamben as an 'activist', *pro*-Negri), biopolitics is set up as the *opposite* of political sovereignty, rather than its most acute expression. Thus, by setting the death camp in opposition to the history of political sovereignty, Esposito not only trivialises the Nazi extermination plan (mere 'normativization'), but simultaneously restores its sacral aura of unique, unfathomable 'Evil', suitable not for historical analysis but only for pious contemplation. (But then again maybe that's the same thing as trivialising it.)

Present day business-as-usual also gets off lightly, thanks to the mealy-mouthed reformulation of 'the camp as *nomos* of the modern' as 'Nazism is the threshold of the contemporary age'. Whereas the first version implicates every ordinary survival transaction in the logic of sovereign death-decision, the second easily accommodates the kind of Sunday supplement Ethics in which *residues* of radical, anomalous Evil have never quite been purged from the global 'community'. (Genocide, as seen at Auschwitz, coming soon to an African civil war nowhere near you!) As Empson notes, the 'bioethical parables' Esposito dwells on are catastrophic (and for the most part exotic, not to say Asiatic) rather than quotidian, much less economic: mass rape in Rwanda, police massacre in Chechnya, China's 'one child policy', or, closer to the European *Heimat*, a French child suing for having missed out on being aborted.

At an oblique relation to some of these questions stands Mark Crinson's review of In the Desert of Modernity, last year's exhibition on colonial (mostly Moroccan) architectural modernism at the Berlin Haus der Kulturen der Welt. The curatorial story counterposes a colonial modernist (which some might pronounce 'humanist', meaning it as a compliment – see J.J. Charlesworth, 'Any Other But Our Selves', *Mute*, Vol 2 #10) architectural utopia to indigenous traditions of *bidonville* 'habitat'-building which occasionally influence the modernist showpieces themselves, before the whole thing blows up in latter-day riots of the former colonised on Parisian modernist estates. Crinson gives short shrift to this compounding of unhistorical myths, noting the marginal status of modernist housing projects in the wider system of Moroccan urban planning 'apartheid', the doubtfulness of the *bidonville* 'influence' on the modernists and the spuriousness of the equation '*bidonville*=anti-colonial revolt'. It could be added that the *banlieues* of the 2005 uprising are not 'Arab' citadels but working class areas where, among other proletarians, a lot of North Africans and their French-born children live. In the present context, though, it's the *bidonville*-fan architects' use of the word 'habitat' that catches the eye: what could be more in the 'nature' of 'bare' African labouring life than such an *animal* form-of-dwelling?

Meanwhile, in Mickey Rourke Virno's all-in bout with 'essential Hobbesian darkness', the fighting professor tries a bold move, an argument he no doubt imagines will have dialecticians spinning where they lie under the Landfill of History.

Linguistic negativity, he says in not quite so few words, is the uniquely human 'condition for' *fascism*, and *ergo*, of course, for 'radical evil'. But rather than inviting Virno's piledriver irony by *negating* this, let us leave him to his ethics of eternal mimicry, sorry, 'simulatory identification' and 'intraspecies empathy'. For this issue of *Mute* also contains Howard Slater's reading of the much less moralistic (and not coincidentally, less biologised) invocation of human negativity in *Species Being and Other Stories* by Frére [*sic*] Dupont. 'The brother' (*pace* na gCopaleen) stages 'an autopsy of disillusionment' in 'the pro-revolutionary milieu', starting from 'revulsion' at the sight of his own 'bad faith'-infested writing. This 'negative', then, is reflexive and vertiginous, and far removed from Virno's fable of the self-*affirmation* of the camp guard who decides on the 'non-humanity' of the 'old Jewish man'. Dupont is hardly the first 'pro-revolutionary' who fearlessly embraced fearful self-laceration. But for the brother this is no dead-end paroxysm of personal despair: Slater sees the bearing of the gesture on Marx's 'species being', understood as 'ongoing lived antagonism between drives and the adequacy, or not, of forms of collective being'. Repudiation of the 'I' (or 'we') is *optimism* for those who stake the future on the self-abolition of their class.

Dupont glimpses something of this at 'messy' mundane level in 'prole reluctance to work': the 'schizoid' (or 'anti-social', or 'self-destructive') refusal of a class to reproduce itself as such. But here he runs into the problems that would follow from universalising 'revolt' (and thus also implicitly its adversary) as the essential content of species being. As Slater puts it: 'you can't sulk a social relation away'. In partial answer, Dupont proposes a 'feral subject' which would 'allow for' a 'pre-human' dimension of affective trauma. Yet if anything this only leads to more problems. The brother's unconcernedness with a 'correct' micro-specialist reading of Marx is welcome, but Slater hears in the 'pre-human' at least an echo of the 'archaic', or maybe of 'primitive communism'. If that echo is there, then 'feral subjectivity' might only serve to weaken the 'species being' of the *1844 Manuscripts* where it is strongest: in the conception of human species-specificity as *nothing but* conscious existence in the continuum of the contingent, 'messy', historical present. It's precisely the *narrowness* of this conception that keeps it open to the traumatic dynamics of collectively lived time, making it far more powerful than any biological and/or spiritual image of 'human nature', including one like Virno's that tries to creep in through the historical back door.

Matthew Hyland is a contributing editor to *Mute*

The MIT Press
http://mitpress.mit.edu

Fitzroy House, 11 Chenies Street, London WC1E 7EY
tel: 020 7306 0603 • orders: 01243 779 777

White Heat Cold Logic
British Computer Art 1960-1980
EDITED BY PAUL BROWN, CHARLIE GERE, NICHOLAS LAMBERT AND CATHERINE MASON

"Here is a fascinating crash course in the early history of computer art seen through the eyes of its pioneers and heirs. This is a living saga of digital innovation pushing against the limits of new technologies and cultural norms. The focus on Britain provides a vital corrective to those who think it all begins in Silicon Valley and MIT."
— **Richard Coyne**, University of Edinburgh, and author of *Cornucopia Limited*

£28.95 • cloth • 500 pp. (63 illus.) • 978-0-262-02653-6

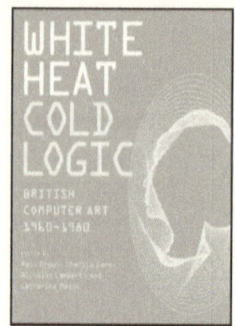

Racing the Beam
The Atari Video Computer System
NICK MONTFORT AND IAN BOGOST

"*Racing the Beam* presents not just the technical challenges but the financial, bureaucratic, and scheduling considerations that harried the Atari 2600 VCS programmers. Modern game designers should read this book for the same reason that modern generals study the military campaigns of Alexander and Caesar: the technology is completely different but the principles are the same."
— **Chris Crawford**, former head of Atari's Games Research Group, and co-founder of Storytron

£14.95 • cloth • 192 pp. (22 illus.) • 978-0-262-01257-7

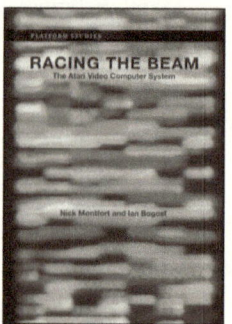

Hijacking Sustainability
ADRIAN PARR

"None of us can afford to ignore sustainability today since the very life of the planet is at stake. And yet it is easy to forget that sustainability is a political problem and a cultural problem too. Hijacking Sustainability is a timely reminder that sustainability is not something we should leave to the market to sort out. Parr makes clear that sustainability is a matter for which we all have to take responsibility and that to do that we have to wake up to what's really going on. Critical theory can scarcely have hoped for a more important book." — **Ian Buchanan**, Cardiff University

£16.95 • cloth • 221 pp. (2 illus.) • 978-0-262-01306-2

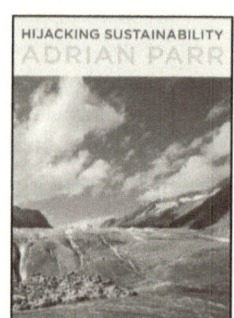

Distributed for Semiotext(e)
Correspondence
The Foundation of the Situationist International (June 1957-August 1960)
GUY DEBORD

TRANSLATED BY STUART KENDALL

Tracing the dynamic first years of the Situationist International movement, Debord's letters — published here for the first time in English — provide a fascinating insider's view of just how this seemingly disorganised group, drifting around a newly consumerised Paris, became one of the most defining cultural movements of the twentieth century. Brilliantly conceived, this collection of letters offers the best available introduction to the Situationist International movement by detailing, through original documents, how the group formed and defined its cultural mission.

400 pp. • £12.95 • paper • 978-1-58435-055-2
£35.95 • cloth • 978-1-58435-063-7

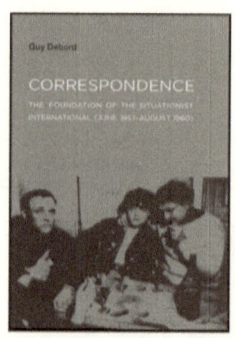

Climate for Change

13 March – 31 May 2009

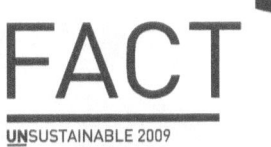

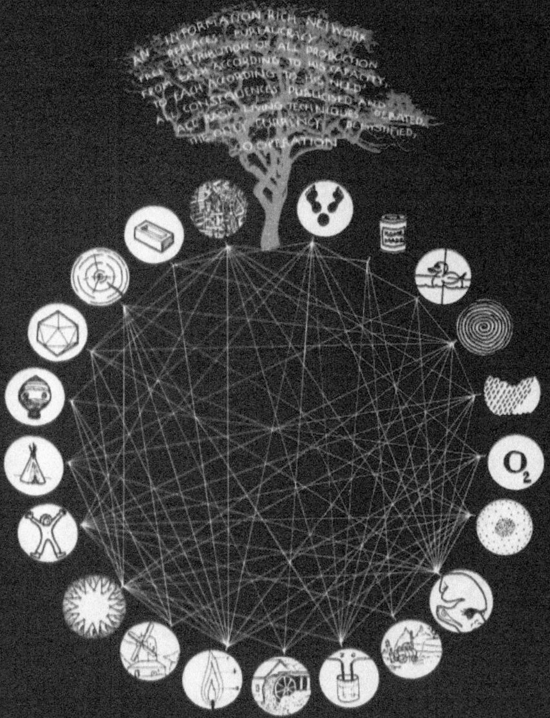

Climate for Change is a unique experiment in activism, engagement and networking, examining the multiple crises affecting the planet – from ecological to financial, food to housing. Featuring Stefan Szczelkun, Eyebeam's Sustainability Road Show, Melanie Gilligan, The Ghana Think Tank Project, N55 and more.

FACT acknowledges the support of Eyebeam Art & Technology Center and Radiator Festival.

FACT (Foundation for Art and Creative Technology)
88 Wood Street, Liverpool L1 4DQ   www.fact.co.uk

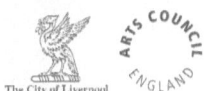

Fourteen years in a nutshell!
A new book about Mute...

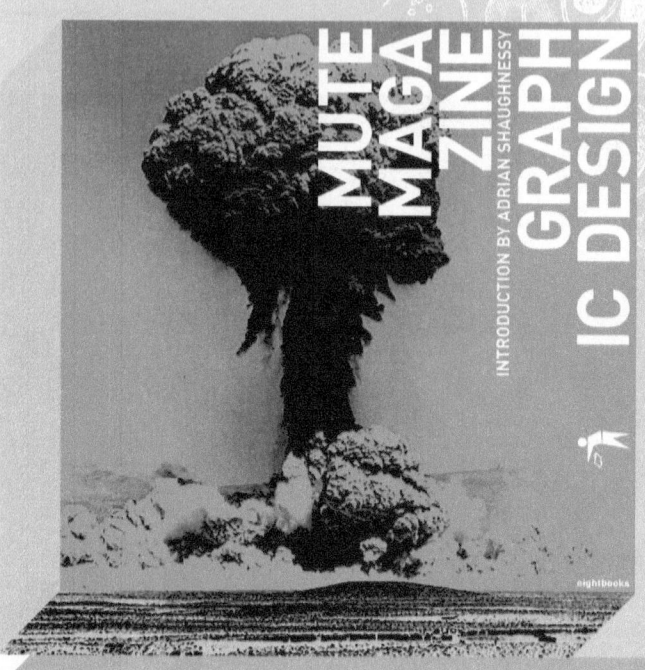

Mute Magazine: Graphic Design (April, 2008)

In the early 1990s, long before the internet became an integral part of life, a handful of pioneering magazines took it upon themselves to imagine the web into existence. Using fiction, interviews, speculative theory and experimental graphic design, London-based Mute wielded an influence disproportionate to its scale. Nearly fifteen years after its launch in November 1994, Mute's publication history defines an era, telling the fascinating tale of one publisher's relationship with the 'digital revolution'. This graphic design history presents a full overview of Mute's output, including logos, covers and spreads.

Introduction by Adrian Shaughnessy, with further contributions from Damian Jaques, Pauline van Mourik Broekman and Simon Worthington.

Published by 8books and now taking orders at Metamute.org/mutegraphics

Softback 220 x 220 mm, 144 pages, 250 colour Illustrations

10% discount for Mute subscribers

Buy it online at:
metamute.org/mutegraphics

UK £ 19.95	Europe €25
US $ 35	ROW €25

The long-awaited
MUTE ANTHOLOGY
out this May!

Proud to Be Flesh: A Mute Magazine Anthology of Culture and Politics After the Net
Eds. Pauline van Mourik Broekman & Josephine Berry Slater

Mute magazine was born, somewhere between art school anomie and the thrill of the World Wide Web's appearance, in 1994. Looking back, Mute's most unchanging feature is its wilful eclecticism and ceaseless criticality. Five years in the making, Proud to Be Flesh brings together the best articles to grace the pages of Mute in a single compendium. Featuring seminal writing on the evolution of Web 2.0, the knowledge commons, new media art, the politics of globalisation, and the material and immaterial ramifications of the net, Proud to Be Flesh is a one-stop reference to the cultural politics of the digital revolution. Thematically contextualised, this is an ideal resource for research, teaching, students, and readers craving a sustained and persistent analysis of 'culture and politics after the net'.

Chapter Titles:

1. Direct Democracy and its Demons: Web 1.0 to Web 2.0
2. Net Art to Conceptual Art and Back
3. I, Cyborg: Reinventing the Human
4. Of Commoners and Criminals
5. Organising Horizontally
6. Assuming the Position: Art and/Against Business
7. Under the Net – the City and the Camp
8. Reality Check: Class and Immaterial Labour
9. The Open Work

Available for pre-order in a hard cover, full-colour, ultra-limited edition.
Relive the best writing, graphics, and design of Mute
624 pages, 48 pages of illustrations
Pre-order price: £45.99 + p&p
Pre-order at: **metamute.org/proud_to_be_flesh**

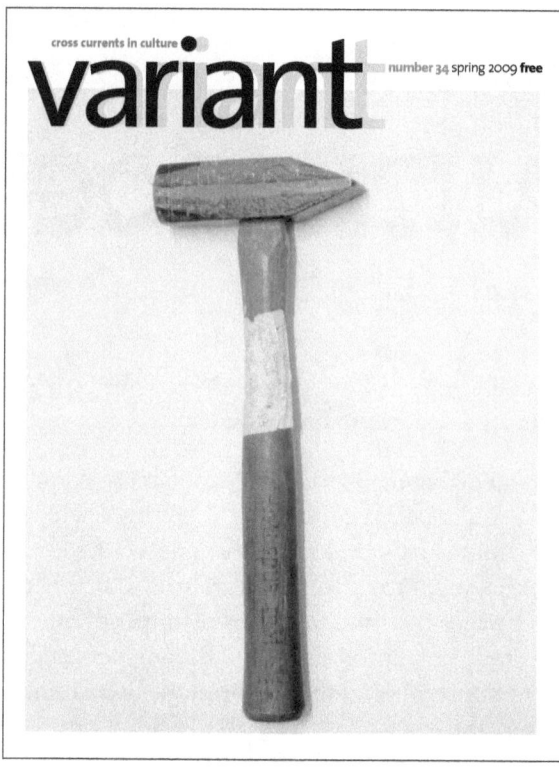

Culture in its broader social political context.

www.variant.org.uk

more is more

Independent media distribution & video screening network

More is More is an open source, online distribution system for small-scale and independent media. The aim of the network is to provide independent media producers and cultural organisations with a platform that can connect them to local outlets and events. More is More facilitates the sale of goods at such locations as well as direct through the website itself.

Commercial distributors are not best geared to the distribution of media products from the cultural, non-profit or political sectors. OpenMute's distribution network is an attempt to develop an alternative. More is More distributes the following: video, magazines, books, comics, posters, flyers and music. It is also possible to arrange your own event or film-screening through the platform.

While the site is at an alpha stage, we are looking for reactions and input from individual media-producers, cultural and activist organisations as well as a variety of outlets that might be interested in putting their products online or selling them locally.

An OpenMute project
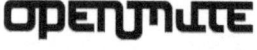

Supported by Digital Pioneers
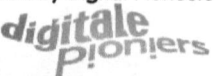

moreismore.net

METAMUTE

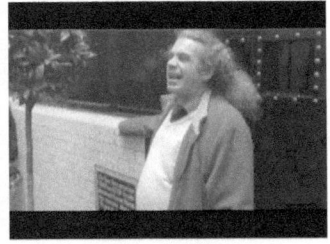

Video - Forever Blowing Bubbles: A Walking Tour with Peter Linebaugh and Fabian Tompsett (2008)
http://www.metamute.org/en/content/video_forever_bl owing_bubbles_a_walking_tour_with_peter_linebaugh_ and_fabian_tompsett_2008

A walking tour and talk in the City of London, taking in landmarks of capitalist crisis past and present. Writer Fabian Tompsett and Historian Peter Linebaugh gave a tour around the City relating the contemporary financial crisis to those of previous eras (such as the 1720 South Sea Bubble), using the urban fabric as text.

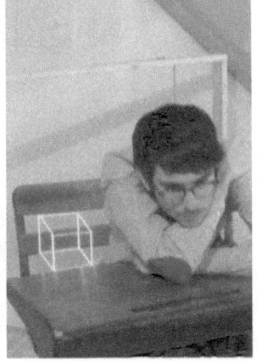

Code Dreams are Made of This
by M. Beatrice Fazi
http://www.metamute.org/en/content/code_dreams_are_ made_of_this

This year's Piksel festival celebrating 'Code Dreams' saw the boundaries between artists, audience, hardware and software blur in the collective pursuit of a machinic unconscious, as well as a highly conscious celebration of FLOSS culture.

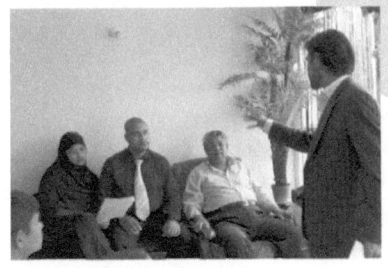

Debt: The First Five Thousand Years
by David Graeber
http://www.metamute.org/en/content/ debt_the_first_five_thousand_years

Anthropologist David Graeber argues that it is only with a general historical understanding of debt and its relationship to violence that we can begin to appreciate our emerging epoch. Here he begins to fill in our historical knowledge gap.

Brutalist Soap
by Kate Rich
http://www.metamute.org/en/content/brutalist_premolition

What lies beyond the failed utopias of the modernist welfare state and the free market? Gail Pickering's recent film/performance, despite its strictly internal focus on life inside a Brutalist housing estate, opens up scope for speculation.

http://metamute.org

BURDENED BY THE ABSENCE OF THE BILLIONS?

> Marx's concept of 'species being' is for some a way of re-connecting with fertile currents in the communist left. **Howard Slater** explores Frére Dupont's recent book *Species Being and Other Stories* as a vehicle of exodus from left orthodoxies

But this negation carries within it a yes which is greater than itself
– Octavio Paz

Over the past few years several publications have surfaced from what can loosely be called the non-Bolshevik revolutionary milieus. Ordinarily publications from such milieus can hardly be noted for their personal openness, play with form and stalwart exasperation with the seeming shrinkage of their circles. Such books as *Call*, *Zones Of Proletarian Development* (ZPD) and this one by Frére Dupont are noteworthy in that they seek, non-prescriptively, to provide grounds for optimism and fresh angles of approach for those milieus that will not rush to embrace them. A provocative theme in their approaches is the way that each reflects upon the modes of organisation of those milieus. Each has experimented with 'phantom organisations' – imaginary groupings of one or several that offer some means of conceptual secession, some means of supported self-exile from those hermetic orthodoxies for whom counter-cultural activists are, as 'culturalists', not to be taken seriously. From *Call*'s elaboration of a party of secession through to Mastaneh Shah-Shuja's investigation of 'reflexive joint activity' in a ZPD, could it be that these books appeal to those distanced from the vestigially workerist revolutionary milieus, or to those convinced that capital's efficacy is, to some degree, related to its instauration as a social relation? Are such approaches, with their accent upon relational congruence rather than ideological purity, more attractive and less threatening for those put off by the over erudite, the emotionally inarticulate and the suicidal militancy that non-revolutionary 'others'

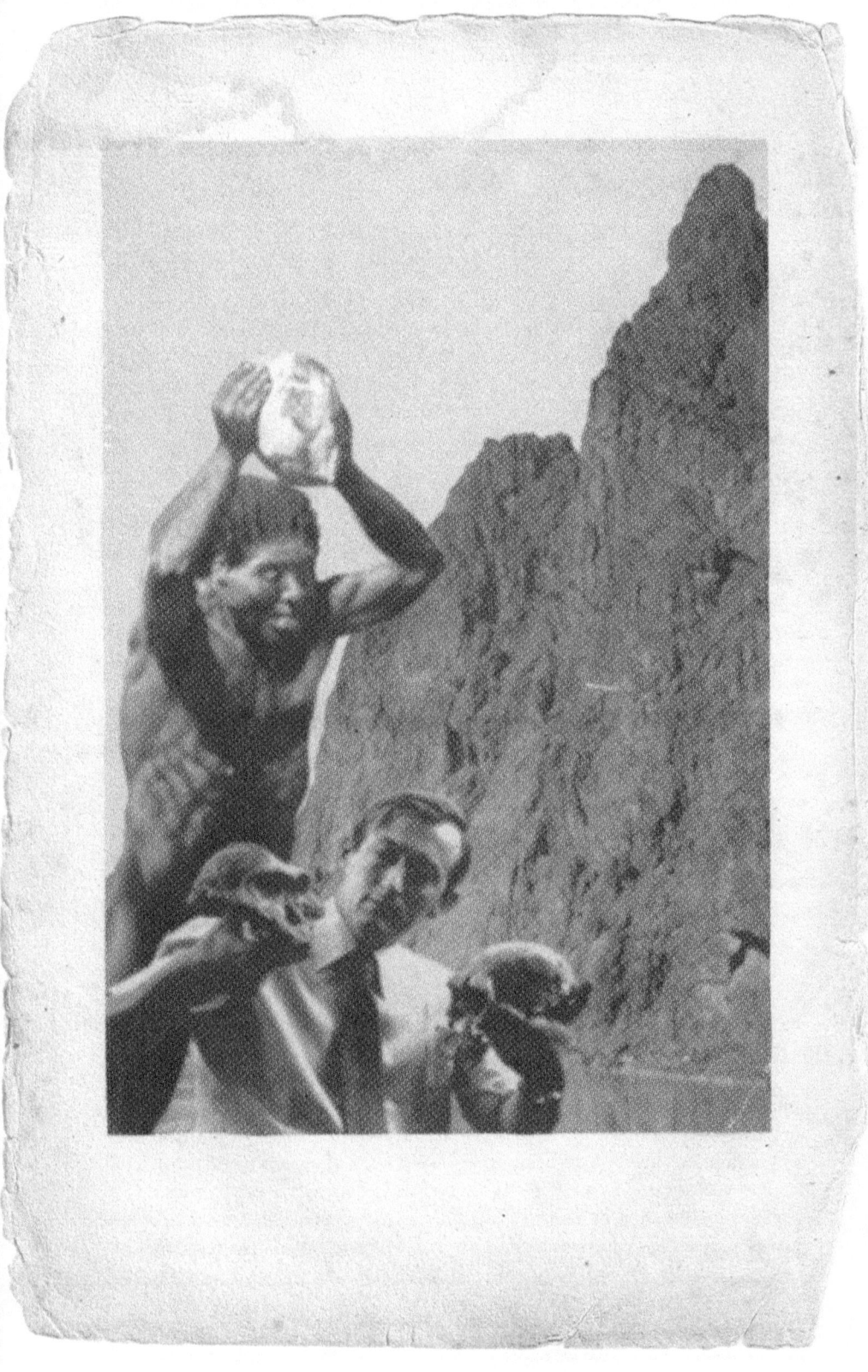

Image: Theo Michaels, *Pre-Human Kills Sceptic*

complain of? Frére Dupont frankly asks the question: 'why is it that others feel no interest for us?' (p.39).

Of these books, it is *Species Being and Other Stories* that could be described as the most personally exacting of the bunch. Being, in some ways, an account and autopsy of disillusionment, it has a self-interrogatory rigour that reminds me of Foucault's insistence that revolutionaries of all hues ask themselves why they identify as revolutionaries: too often the 'role' is taken for granted. For Frére Dupont it could be more a matter of talking of a 'pro-revolutionary milieu', of getting rid of the identitarian baggage, moral purity and dysfunctional personal relations that abound, and embracing, instead, a manner of being that befits what he calls the 'for-human' of species being: 'only when the left despairs of itself is there room for a vaguely human becoming' (p.102).

That brother Dupont does not intensively interrogate, in an explicatory way, what Marx means by species being, but offers it up as a non-foreclosed 'for-human' (with all the pitfalls that could entail), means that as readers we must suspend our yearning for 'received truth' and participate in the suspension of certainties. Indeed, there is nothing certain in what Marx says: 'Conscious life activity directly distinguishes man from animal life activity. Only because of that is he a species being.'[1] But such conscious life activity does not necessarily mean theoretical work, but maybe means an ongoing lived antagonism between drives and the adequacy, or not, of forms of collective being. The presence within us of these drives and affects that cut across and cut up our rationalising and forecasting is, for me at least, an element of species being that persists as a common human trait that ideological production cannot appease. An articulation of the messiness of all this is a possible take on Dupont's 'for-human', and it is a messiness that Dupont does not recoil from:

> I cannot bear to face what I have written – bad faith dogs me. I have scanned through the words of course [...] That was more than enough to fill me with revulsion. (p.iii)

Revulsion, bad faith, the fear of attack and of finical criticism. Is this what it's come to for us?

This 'revulsion' endears Frére Dupont to me, and I hope in passing to take up its umbilical thread after outlining a little the eclectic content of this *sui generis* book. A book that whilst riffing, in part, on such currently debated concepts/practices as 'communisation', manages to come across like a work of experimental prose. The title *Species Being and Other Stories* is as good a place to start as any. Brother Dupont's mobilisation of a sense of fiction, in what is ostensibly a work of theory, enables a refreshing candidness and gives free reign to speculative and non-fore-

Howard Slater

Revulsion, bad faith and the fear of attack. Is this what it's come to for us?

closing flights (speculation and playfulness, and their attendant 'messiness', being often misunderstood and a cause for barbed comment in the milieus?). It enables brother Dupont to 'begin again from a slightly different position' (p.16) and, as already suggested, to add his own take on the meaning of 'species being'. That Marx's *1844 Manuscripts* are a touchstone for this book should not be overstated since Frére Dupont improvises with this hazardous and dimly extrapolated phrase. Indeed, for me, it could be said that species being is not to be taken as a past state that we strive to reattain, some sort of human essence, but a work of collective (fictional) production; a malleability of seemingly innate ahistorical drives. For others, such as the group around *Internationalist Perspective*, an interesting angle is taken in their setting species being and social being in opposition/conjunction. And then there are those for whom the *1844 Manuscripts* and musings around them are a cause of shame and revulsion at the thought of a 'humanist' and pre-scientific Marx. But, as this brother seems to show, there is, in the 'terror of the dawn chorus', in the value-imposed balance sheet of a life, always the sense of a 'becoming human' to clutch to.

In some ways Frére Dupont's take on species being as the 'for-human' and the 'pre-human' is his tabula rasa, his beginning again from a different position, his attempt to find some 'invariants' or common human traits in the ongoing struggle against capitalism. And, who knows, to find a conceptual space for the working class after it has abolished itself! As with Jacques Camatte, the turn to species being here is tied up with a sense of the formation of a 'human community' as the overarching communist tendency. A tendency that is counter to the reproduction of the species being as labour power that this brother asserts is, in the absence of capital investment and with a prole reluctance to work, being left to the liberal state: 'the working class constantly prepares itself for its return to species being, seeking its own level through this implied rejection of itself as working class' (p.70). This failure of reproduction as labour power as well as the schizoid position of a class subject that, so the theory goes, needs to desire its own dissolution, opens up notions of whether the working class is still the revolutionary subject. I get the sense that for this brother it is the milieus that are the blockage in the revolutionary process; that and the ideological mediation (encapsulated in organisational forms) that form a barrier between them and the Others.

That this brother asserts that the 'for-human' is a more common form of being than the role of the revolutionary implies, I think, that for him the revolutionaries are blind to the innate revolt of the 'for-human'; a revolt expressed not as a 'political use-value', but as a means to 'do everything to keep and increase our dignity as living things'.**2** That species being is here made the synonym of revolt, and that revolt is made the 'essence which every human may access' (p.65) may give grounds for a much needed 'return to optimism'. However, it clashes with a leading question of the book: 'why does the proletariat not revolt against its conditions?' (p.viii). In response to his own musing Frére Dupont says that these Others aren't inspired, that we in the milieu are solipsistic, have an insular self-regard, feel the individualised pressure of auto-culpabilisation and are increasingly negative. It's tempting to also suggest that the milieus (in all those separate waiting rooms) are so off-putting to the Others because each room feels, somewhat pathologically, that they are in 'possession' of the correct consciousness, the correct analysis, the skeleton key. Indeed, as a refreshing counterpoise to such righteous foreclosure, the meandering, engaging and almost extemporising aura of this book sees Frére Dupont later stating with nonchalant assertiveness: 'It is never a matter of revolt becoming the vehicle of a solution' (p.75)! Revolt is not enough; you can't sulk a social relation away.

Revolt is not enough; you can't sulk a social relation away

Whilst it is against the spirit of this eclectic and engaging book to attempt to place its 'conjectured ground' within a rational framework (critical appropriation), I do struggle a little with the demarcation of the 'pre-human' element of Frére Dupont's take on species being. It seems to function on too many levels at the same time: as an exploration of the neolithic (or 'primitive communism'?); as a reference point to a state of precarity that is aligned to a human condition and not something of recent invention; as a harbinger (or childhood memory) of a relational existence unencumbered by the abstractions of value; as the persistence of the irrational and of a surplus; of the founding of human societies around death and ritual. The 'pre-human', then, is more succinctly offered up as an explanation to the 'destructive character of small group psychologies' (p.33) in that Frére Dupont writes 'up to this moment groups have tended to allow the existence of an untheorised pre-human element hostile to their own expressed values' (p.23). Whilst the many surrealist organisations and group psychotherapy practitioners who work with the 'untheorised' and with the 'hidden third' of relational dynamics may feel a bit miffed to be overlooked, I can only think that the 'pre-human', in this context, is the unspoken elements of unconscious life; the transferences, projections and sublimations associated with group life; the persistences of affects that circulate between us; the clash between these mute feelings in search of words; and those 'expressed values' too full of the trickery of ideological languages.

Frére Dupont puts forward the tentative suggestion that pro-revolutionary groupings (already the prefix 'pro' is freeing for us Others) should embrace the 'pre-human', or, as my baggage dictates, unconscious group dynamics, in order, I reckon, to more roundedly embrace the becoming inherent in the 'pro-human'. The two, for Frére Dupont, are in interrelation and this is given outline in the section entitled 'We Build Complex Assemblages' in which he extrapolates on a phantom organisation called earthen cup that has as its platform 'the untheorised and non-included aspects of human existence'. In this section of the book Frére Dupont sketches out an 'organisation for those who have no organisation' (Bataille), an immanent organisation, an 'associative medium' (c.f. surrealist groups), that rings out with both a poetic yearning and a declaratory tone that is vaguely self-mocking. He rounds off this nine point anti-manifesto as follows:

> Our purpose is to develop a feral subject, that which even if it appears under present circumstances, is actually determined out of time, by both the most ancient past and the most distant future. The subject earthen cup seeks to invoke has its hands upon the levers of its own transformations, its mouth issues a code of metamorphoses. (p.47)

The feral subject invoked here has echoes of Antonin Artaud (cited earlier in the book), as does the persisting notion that earthen cup come to embody some form of expanded theatre and that pro-revolutionaries engage in role playing games.[3] The spirit of the Bataille of *Acephale* and the Artaud of *Theatre of Cruelty* peek in here especially in their similar insistence and interest in summoning up the beyond rational (discourse) of 'primal' elements.[4] Such a 'beyond rational' also surfaces in Dupont's 'feral subject' being 'determined out of time'. But rather than this having to be seen as some wildly transcendent subject it could be read, rather, as the furthest extension of a ZPD: the persistent proximity within us of the archaic and the modern; a marker of species being.

Artaud, then, provides something of a prism for me in the context of this book; a prism that also allows us to catch a glint of Marx. It's a simple phrase but, as cited by Frére Dupont, Artaud, in talking of his prospective theatre (a combination of the pre-human and for-human, the archaic and the modern?), says that its efficacy is in 'compelling men to see themselves as they are' (p.25). This may seem like nothing to revolutionary identitarians distanced from the Others, but its implication is that we can only see ourselves as who we (temporarily) are by drawing upon the 'untheorised and non-included aspects' of ourselves; the affective elements that seem not just superfluous to the grand theoretical discourses that we immerse ourselves in, but seem like indications of our own revolutionary inappropriateness, our revulsion in thinking in knowing-all tones, our shame at being determined bourgeois subjects. Frére Dupont: 'everything existent under the capitalist conditions transports value for the economy' (p.63). If we, to misapply Marx, are 'independent centres of circulation', if the makeshift formula 'I=value' holds, then there is a compelling case to see ourselves as 'who we are', as who we have been produced to be, in order to partake in the process of becoming human, of becoming species being.[5] This is itself part of the struggle against capitalism that remains 'non-included' as such. Rosi Bradiotti has stated it starkly thus: 'one has to contemplate the unedifying spectacle of one's own failings or shortcomings.'[6]

> # suffering and trauma are human constants that no revolution can eradicate

Such a 'going fragile' is no easy task and it is to be wondered whether by 'cruelty' Artaud meant that a kind of 'autotraumatisation' was called to be delivered up by his proposed receptor-participants. This perhaps leads us to the Marx of the *1844 Manuscripts* and another take on species being that, again, could be called an 'invariant'; namely Marx's assertion that,

> man as an objective sensuous being is therefore a suffering being, and because he feels his suffering, he is a passionate being. Passion is man's essential power.[7]

Passion doesn't really go out of fashion (well, not yet), and suffering and trauma are human constants that no revolution can completely eradicate. In brother Dupont's terms, passion and suffering are both 'pre-human' and 'pro-human'. In a touching passage he urges that pro-revolutionaries 'invite others to reflect upon the truth of their own personal anguish, and thereby recognise their relation to the world' (p.69). Such an attunement to their 'own feelings of revulsion for the organisation of the

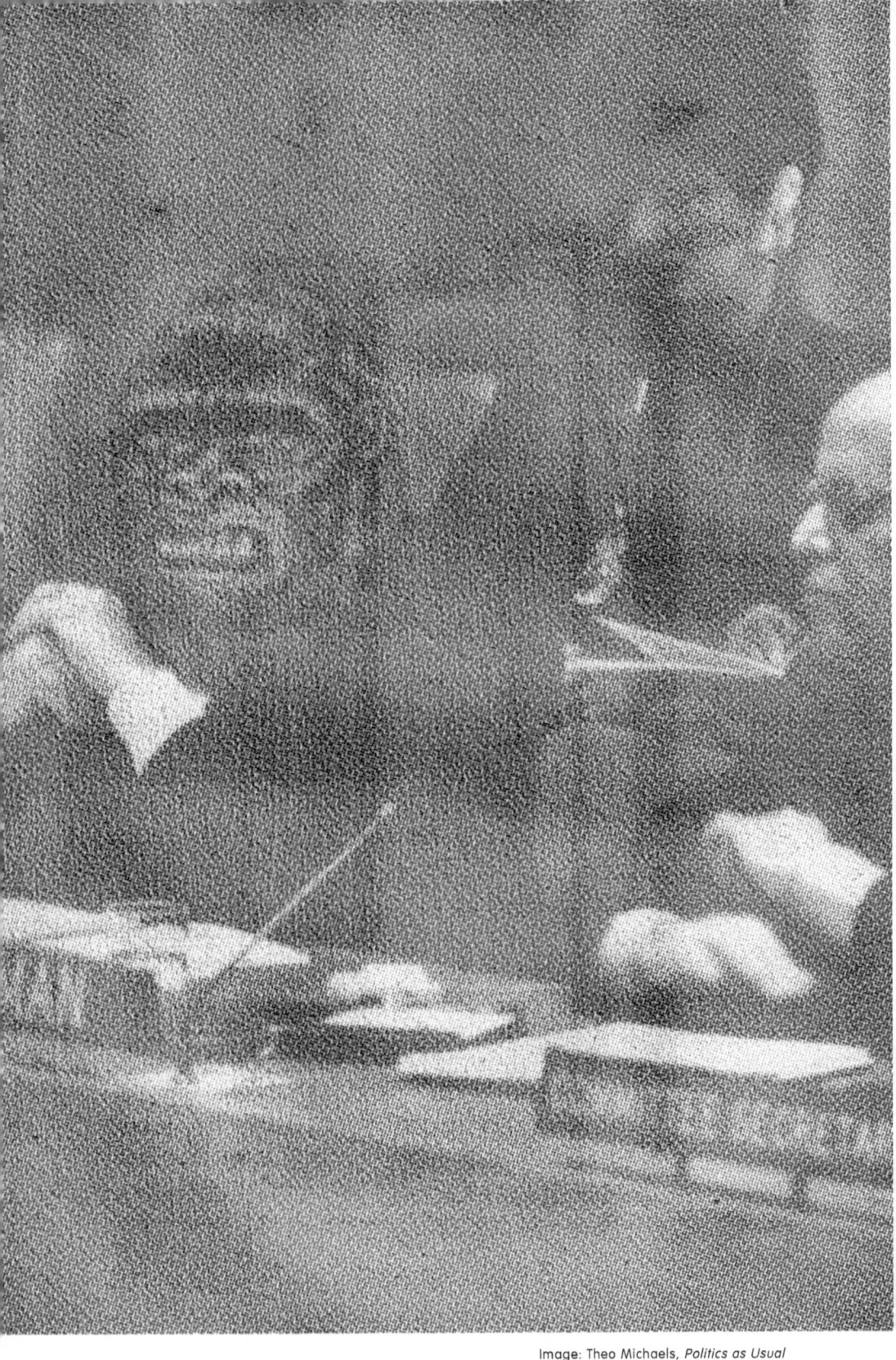

Image: Theo Michaels, *Politics as Usual*

world' may sound close to counselling or psychotherapy, but these latter are, on the whole, bogged down in the individualising 'therapeutic dyad'.[8] Instead brother Dupont, in 'Letter To T', returning to his organisational musings, offers crucially that 'we must visit our frailties into the context' (p.118). Does he mean that our frailties, our emotional susceptibilities, our suffering, our blocks, our confusions, our fantasies, our feelings of alienation etc., should be allowed into the pro-revolutionary group context and be seen there as indications of our endocolonisation by capitalist valorisation imperatives and not as our unworthiness to the cause? Frére Dupont: 'we are dependent on mediated forms; our subjectivity echoes, even desires, the reproduction of these forms' (p.105). If this sounds to some like a return to Encounter groups, Maoist self-flagellation sessions or mid '70s pro-situ auto-crucifixion, then, it could be countered that these were all ideological practices that actively repressed the 'primal' or 'pre-human' element of species being in favour of the self-protecting sleights of discourse/power. Foucault:

> the manifest discourse [...] is really no more than the repressive presence of what it does not say; and this 'not said' is a hollow that undermines from within all that is said.[9]

To allow for and work with the 'not said', the affective residue, gives an insight into a compelling nuance of what we are fighting against: capitalist valorisation imperatives as productive of a 'paranoid-narcissistic ego'?

If this all sounds very individualistic then we must remember all those revolutionaries suicided by society as well as the personal costs of commitment and its deliberate precarity that Frére Dupont touches upon in the pages of his book. It is this level of psychical suffering that goes mostly unnoticed in those milieus that reduce their subject to the condition of the physically and mentally exhausted labourer. This says nothing about the levels of despair that follow upon protracted exhaustion nor does it give credence to the potential intensification of such suffering for those in the milieus: you're exploited, you know it, and worse you feel it, but you cannot manifest these feelings in the prevailing language of the milieus. If 'I=value', if value is breath, then, in the depths of a psychical suffering you can become intra-alienated; the relations between your own modalities breakdown in the generalised equivalent of a 'self' that cannot become, that cannot risk its own difference. This makes us ripe not only to conform to the social situation as it is, but to submit to the 'unmessy' orthodoxies of the revolutionary milieus for the want of anything more effectively and affectively engaging on a variety of 'pro-human' levels.

And yet, there is, I feel, something pulsing and inchoate that is being brought slowly to expression in both this book and the others cited at the beginning of this review-article. Before drawing upon these sources it is useful to consider again the Marx quote above. Here, albeit before he embarks on his critique of political economy, Marx does not reduce the species being to 'labour', but to passion and suffering, i.e. an opening out of species being and not a reduction to one of its many (potential) facets. So what happens if we couple this to the phrase 'affective classes' reportedly uttered by Walter Benjamin during a trip to Paris in 1935? Amidst all the taxonomies and re-writes of class as a precariat, multitude, entrepreneuriat etc., this phrase of Benjamin's remains for me the most provocative of the lot. What did he mean by this phrase that he didn't explore for himself but which was reported and glossed for us by Pierre Klossowski? Was he outlining a 'phantom class'?[10] With this in mind, and inspired by the speculative tenor of Frére Dupont, yet with my own feelings of revulsion at adding to the list, could it be offered that the working classes are being recomposed as affective classes? What would this entail? A Fourierist notion of eroticised and pleasurable work? An appeal to the cracked-up to unite? A suggested point of convergence around a species-activity informed by desiring-production? A re-appropriation of affective labour? A more adequate response to a real subsumption marked by biopower? Does Benjamin's phrase, then, have any relation to *Call* when the anonymous authors have it that the Party 'could be nothing but this: the formation of sensibility as a force' or elsewhere speak of 'affective circulation'? Does it relate to Jacques Rancière's notion of a 'distribution of the sensible'? And what impact would Jonathan Beller's idea of a 'labour theory of attention' add to it?[11] A great deal of collective thinking/practice (or a good critical lashing) would need to be done in this area, but Frére Dupont speculates in much the same direction himself:

this is no return to Maoist self-flagellation

> for the left this recomposition of struggle into an intimate bodily reaction feels like a retreat but they are wrong [...] Revolt is an intimate relatedness to the world and therefore most real at the level of immediate feeling. (p.68)

So, in the revolt against economically induced suffering, a revolt in which we feel, feel-for and attempt to feel-with (empathy), we must also remember that these feelings too revolt against their bearers, that feelings are in revolt, that commitment is not guaranteed as circumstances change and the struggle to survive presses 'inwards' and exacerbates individualism and its pathological variants. In these circumstances, forms of relation, of being-with, seem to become of paramount importance; relations which go towards co-creating a culture that is informed by the 'pre-human' (unconscious) and the 'pro-human' (increased dignity of living things), that encourages the mutual disclosure of a 'going fragile' whilst militating against the appropriation for value of our sensual bodies' capacity for suffering and passion (capital makes money out of our death and our exuberant 'vital force'). Affectivity is at stake, the capacity to feel and be impassioned into revolt, to have feeling destabilise our selves enough to risk the making of an 'unnatural' difference, to no longer have shared feelings of revulsion informed by the operations of mass culpabilisation, by the inherent authoritarianism of language, by the fear of an inhuman, value-laden judgementalism of our worth to enterprises, to the state and, sadly, to each other. So, maybe, for some 'pro-revolutionaries' it is necessary to toy with that 'phantom class' of the affective of which Benjamin is still foretelling. Open up front.

Footnotes

1 Karl Marx, *Early Writing*, London: Pelican 1984, p.328. A discussion on species being by some members of the Internationalist Perspective group can be followed at, http://internationalist-perspective.org/IP/ip-archive/ip-archive.html. The exchange was inaugurated by Rose in Issue 43 (2005). Thanks to N for this link.

2 Adapted from an interview by Tatiana Kondratovitch with Pierre Guyotat titled 'Art is What Remains Of History'. See *Frozen Tears*, Vol. 2, 2004. See likewise Franco Beradi saying '... the inhuman appears as the dominant form of human relations'. See his 'Obsession With Identity Fascism' at www.generation-online.org/p/fp_bifo3.htm

3 These role playing games are not a million miles away from Mastaneh Shah-Shuja's notion, developed through a reading of Lev Vygotsky, that a Zone of Proletarian Development can be an 'imaginary scenario where participants within a ZPD could actively reflect and expand on the debate without feeling pressure to enact preconceived roles and positions.' See Mastaneh Shah-Shuja, *Zones of Proletarian Development*, London: Openmute, 2008, p.100.

4 Paolo Virno has touched on this too in the opening essay of his most recently translated book. Here, in talking of exodus, he says that an element in the dignity of exodus is entrusted to a willing confrontation with 'the murmurings, the dangerous instability of our species'. See Paolo Virno, *Multitude: Between Innovation and Negation*, Semiotext(e), 2008.

5 In issue 48 of *Internationalist Perspective* Sander makes the interesting point that 'the cost of production of the workers as a subject of capital, as subjected to the law of value, not through coercion or even the constraints of the need to earn a living, but in his/her consciousness, values, beliefs culture [...] is a complex of issues that Marxism has undertheorised.' I wonder whether a hermetic Marxism is up to this task; a Marxism of the milieus that doesn't seem to take seriously those ramifications of real domination such as 'biopower' and 'the production of subjectivity' that have been developed by Foucault, Guattari, Deleuze and latterly Negri?

6 Rosi Braidotti, *Transpositions*, Cambridge: Polity Press 2006, p.201.

7 Karl Marx, ibid., p.390.

8 Movement away from the individualising dyad and towards a more social and interrelational therapy have always been a staple of group psychotherapy, but many practitioners in the one-to-one counselling sphere, regardless of discipline (i.e. Person Centred, Psychodynamic and Existential psychotherapies) are embracing what's been called the 'socially positioned individual'. See the work of Gillian Proctor, Pete Sanders, Mick Cooper, Peter Schmid, Lewis Aron, Ernesto Spinelli etc. Some of this work is available from PCCS Books. See, http://www.pccs-books.co.uk/

9 Michel Foucault, *Archeology of Knowledge*, London: Routledge, 1995, p.25.

10 'Pierre Klossowski: Entre Marx et Fourier', extracted by Denis Hollier in the appendices of *College of Sociology*, Minneapolis: University of Minnesota Press, 1988, p.389. The phrase 'affective classes' is reported as arising when Benjamin was pressed by members of the College Of Sociology to describe his take on a 'phalansterian revival' (i.e. a reassessment of Fourier's utopian ideas). Klossowski reports: 'Sometimes he talked about it to us as if it were something "esoteric", simultaneously "erotic and artisanal", underlying his explicit Marxist conceptions. Having the means of production in common would permit substituting for the abolished classes a redistribution of society into affective classes. A freed industrial production, instead of mastering affectivity, would expand its forms and organise its exchanges, in the sense that work would be in collusion with lust, and cease to be the other punitive side of the coin.'

11 See *Call* (n.d.); Jacques Rancière's *Politics of Aesthetics*, New York: Continuum, 2004; Jonathan Beller's 'Vertov and the Film of Money' at, http://muse.jhu.edu/journals/boundary/v026/26.3beller.html. Interestingly and in correspondence to some of the themes of this review-article, the anonymous authors of *Call* have suggested the idea of a 'human strike'. This has also been mooted by the Claire Fontaine group. See their text 'Ready-Made Artist & the Human Strike: A Few Clarifications' at http://www.clairefontaine.ws/text.html. Thanks to A for this latter link.

Bibliography

Frére Dupont, *Species Being and Other Stories*, Ardent Press, 2007. Distributed by Little Black Cart. See, http://www.littleblackcart.com

For 'Going Fragile' see, Mattin/Radu Malfatti at, http://www.formedrecords.com/formed03.html

Howard Slater <howard.slater@googlemail.com> is a trainee counsellor and sometime writer who lives in East London

Paolo Virno's latest book contends that the question of human nature – good or evil? – is suddenly topical, thanks to 'immaterial labour'. But, if true, how useful is this insight?, asks Nina Power

HE'S NOT BEYOND GOOD AND EVIL

Seen from a certain angle, the history of political theory is always, and at the same time, a set of claims about human nature. For several decades, however, on each side of the political spectrum, the tendency has been to worry that the very concept is a limitation at best and a mistake at worst – isn't the image of man the very thing that prevents us from thinking properly about structure or process? Didn't Marx only really start talking about capitalism once he'd shrugged off the humanist idealisms of his youth? 'Human nature' seems a clumsy, old fashioned thing, redolent of outdated philosophies and dubious biology.

Paolo Virno, arguably one of the finest thinkers to emerge out of Italian post-workerism, begs to differ, however, in several rather ingenious ways, linking current research on mirror neurons to Aristotle's theses on praxis, Wittgenstein's discussions of rule following to Freud's writings on jokes. But this is no happy account of the curious little wordy biped. On the contrary, for Virno, Homo sapiens is, if anything, constituted entirely negatively: 'the animal whose life is characterised by negation, by the modality of the possible, by regression to the infinite' (p.18). Against the phenomenological idea of the world as a kind of background and source for all our other possibilities, Virno understands the world as a space of natural conflict, a source of perpetual confusion and a constitutive disorientation. If we are to return to a kind of natural political theory, Virno would want us to understand that this is, above all, an *unhappy* naturalism.

In an article from the first issue of the Italian journal he co-founded in 2004, *Forme di Vita*, Virno states that:

> the content of the global movement which ever since the Seattle revolt has occupied (and redefined) the public sphere is nothing less than human nature.[1]

In the first section of *Multitude: Between Innovation and Negation* , he similarly reminds us that:

> it is not wise to turn one's philosophically sophisticated little nose in the face of the crude choice between: 'man is by nature good', and 'man is by nature bad'.

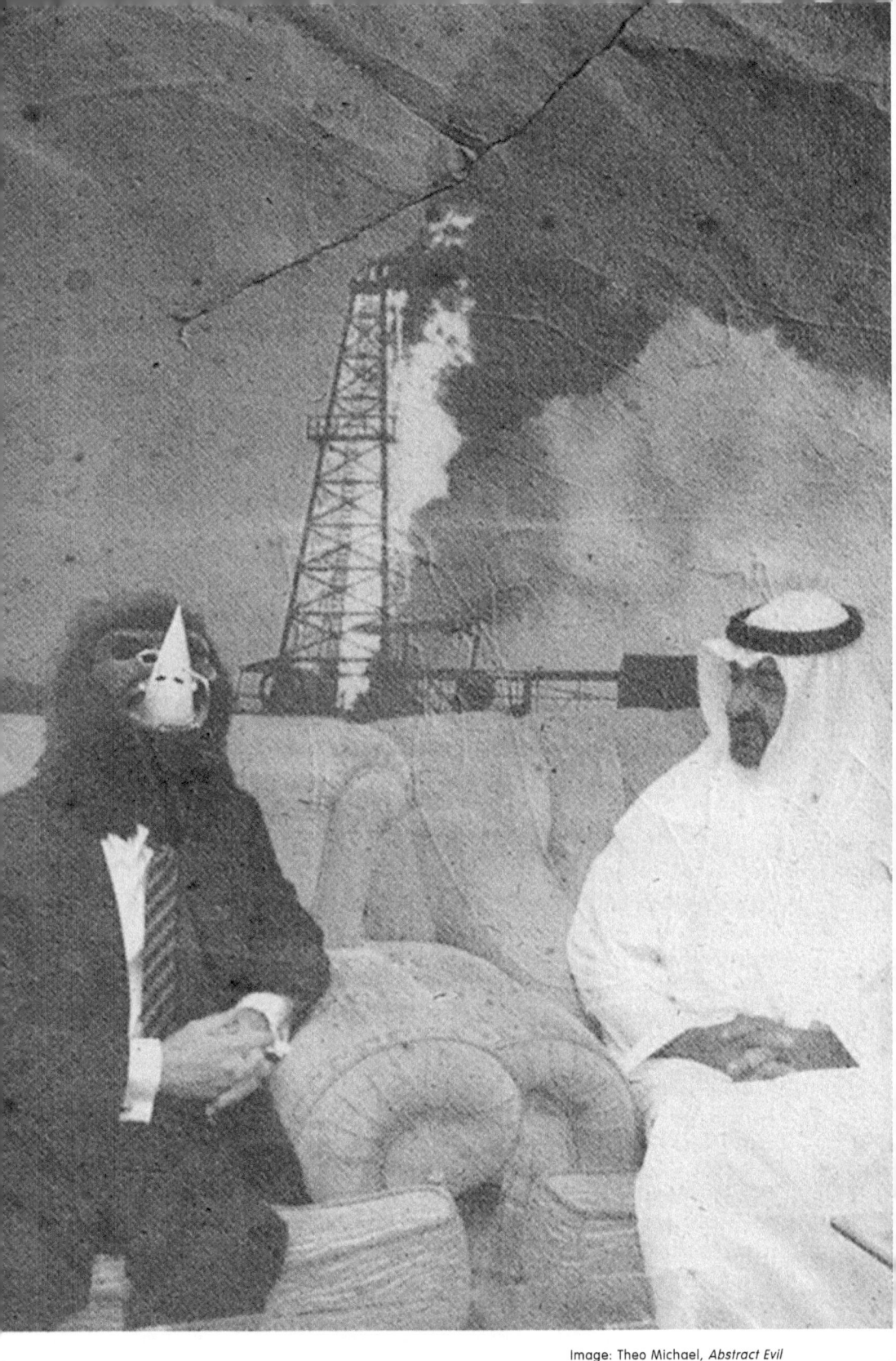

Image: Theo Michael, *Abstract Evil*

It is not, then, in the false choice between Hobbes' *Homo Homini Lupus Est* ('man is a wolf to his fellow man') and an optimistic Rousseauean romanticism that the space of politics lies, but in the 'problematic' temperament of the human animal that is, according to Virno, potentially always 'evil'. Virno turns the questions of classical conservatism into the blueprint for a post-Marxism which understands that the global struggle is motivated by ethics as much as exploitation, by the search for the good life as much as the struggle against bosses.

Unlike the more opaque pronouncements of *A Grammar of the Multitude*, also published in English in 2008, it is in the *Forme di Vita* piece that Virno lays out the explicit stakes of his political project. The distinguishing trait of the movement is

> the extremely tight entanglement between 'always already' (human nature) and the 'just now' (the bio-linguistic capitalism which has followed Fordism and Taylorism).**2**

This, in a nutshell, is Virno's wager – that it is only now, when the differential traits of the species (i.e., that which separates us from other animals, namely verbal thought, the transindividual character of the mind, neoteny, the lack of specialised instincts) are the 'raw material' of capitalist organisation, that we can return again to the question of a politics of human nature. Thus the problem of the 'natural' emerges *contingently*, that is, at a certain historical moment, yet as if for the first time. Virno reminds us of Marx's claim from the *Economic and Philosophical Manuscripts of 1844*:

> It can be seen how the history of *industry* and the *objective* existence of industry as it has developed is the *open* book of the essential powers of man, man's psychology is present in tangible form.**3**

But the difficulty here for Virno is identifying the cracks in the edifice – what separates the exploitation of human capacities under 'biolinguistic capitalism' from the resistance to such forms of exploitation? Michael Hardt and Antonio Negri's rather formalistic apparatus in *Empire* involves a kind of flipping of the switch; if only the multitude could just appropriate their currently exploited cooperative potential, everything would be just as it is, yet, at the same time, completely transformed. Virno's take on the concept of the multitude is subtler, though rather minimal given the title. Presumably the concept sat well with the marketing department.

The contemporary multitude is, then, perhaps the first truly historico-natural being. Or is it? The idea that bio-linguistic capitalism is all that new or all that paradigmatic has been contested in several ways. Firstly by those who argue that immaterial labour still represents only a small portion of total labour, and secondly, by those who argue that the distinction between material and immaterial labour has never been all that clear (and furthermore that the old Taylorist idea that 'you are not paid to think' is far more characteristic of most contemporary labour, immaterial or otherwise, than any exploitation of fundamental human capacities). Steve Wright, among others, has further pointed out that 'affective' labour – denoting those jobs that directly involve care, compassion and kindness (or at least their simulacra), from housework to reproduction to the sex industry – have long accompanied and indeed made possible the labour we associate with more manly, 'proper' work.⁴ The so-called 'feminisation of labour' (the increased participation of women, the centralising of those affects associated with 'women's work') might be a kind of belated recognition of certain labour processes that have been going on for a very long time indeed, not to mention the way in which women's relation to the job market has historically tended to operate. As Silvia Federici notes, simply, 'women always had a precarious relation to waged labor'.⁵

this is no happy account of the curious little wordy biped

Nevertheless, let's accept for the time being that Virno is onto something, that the multitude

> exhibits, in its very mode of being, the peculiar historical situation in which all the distinctive traits of human nature have earned an immediate political relevance. (p.64)

Thus political anthropology takes the place of class struggle by conceding that class war has wormed itself all the way down to the cerebral cortex. It is as if for Marxist reasons – i.e., the shift in the characteristic of the labour process in our mode of production – we must once again become Feuerbachians, concerned far more with generic human capacities than the antagonisms of the working day.

But, unlike Feuerbach, Virno resorts less to a kind of vague humanism than to recent biological research into two main areas relating to human development: neoteny (the retention of formerly juvenile characteristics produced by the retardation of somatic development) and mirror neurons (the neurophysiological phenomenon whereby when we see someone performing an action, the same neurons are activated in the frontal lobe of the observer, demonstrating a kind of original intersubjectivity that precedes the constitution of the individual mind). But rather than take the relative openness of our species and our apparent inability *not* to empathise with others as the basis of political optimism, Virno reminds us that

> every naturalist thinker must acknowledge one given fact: the human animal is capable of *not* recognizing another human animal as being one of its own kind. (p.181)

He's Not Beyond Good and Evil

The problem of evil, with which Virno begins the collection in his discussion of Hobbes, rears its disconcerting head once again: and it is language and its essential negativity that causes the problem. The Nazi camp guard is capable of 'not-recognising' the Jewish captor by the force of the linguistic negative: this is not a man. As Virno argues:

> Negation [...] certainly does not obstruct the activation of mirror neurons; but it renders the signification of these neurons as something ambiguous and reversible. The Nazi officer can consider the old Jewish man to be 'not human', even if he fully understands the old man's emotions by means of simulatory identification. Verbal thought destabilizes intraspecies empathy: in this sense, it creates the condition needed for what Kant has called 'radical evil'. (pp.183-184)

As with all theories of evil, however, there's a danger of reifying it, of turning it into something far beyond the reach of any explanation whatsoever. As much as Virno seeks to distinguish himself from Chomsky's attempt to link the human desire for autonomy to the innate creativity of language and thus avoid any optimistic theory of human nature, he ends up with some horrifically bad examples of evil, using, of all things, the Superdome in New Orleans after Hurricane Katrina to describe the 'oscillation' between the good life and its opposite. One might think that a materialist explanation of the appalling organisation of the evacuation and the racist and classist discrimination towards those who lacked cars and money to leave might be more useful here than any discussion of 'evil'. The appalling events that supposedly took place in the Superdome (rape, murder, mass violence) were, after all, fabricated for a media eager to believe that poor black people are barely civilised, while the very real material constraints of the situation (overcrowding, lack of access to bathrooms) hardly point towards some kind of essential Hobbesian darkness.

Still, one cannot be too enamoured with the world as it is. Unlike the optimism of claims he makes elsewhere about the 'global movement', however, Virno is in these essays much more cautious:

> the multitude is negation, and the negation of negation, the uninhibited 'it is possible that it might be' and the limiting 'it is possible that it might not be'. (p.65)

The problem here is the amount of work done by the related concept of *exodus*. If the production of value is somehow spread across society as a whole, if the working day bleeds into the hedonistic night, if capitalism has managed to monopolise *all* basic human capacities, then the only 'way out' might start looking like a quick dash across the desert. Or, alternatively, like more minor and rare forms of 'innovative action' that break rules, thus indicating the inherent transformability of all linguistic games. After

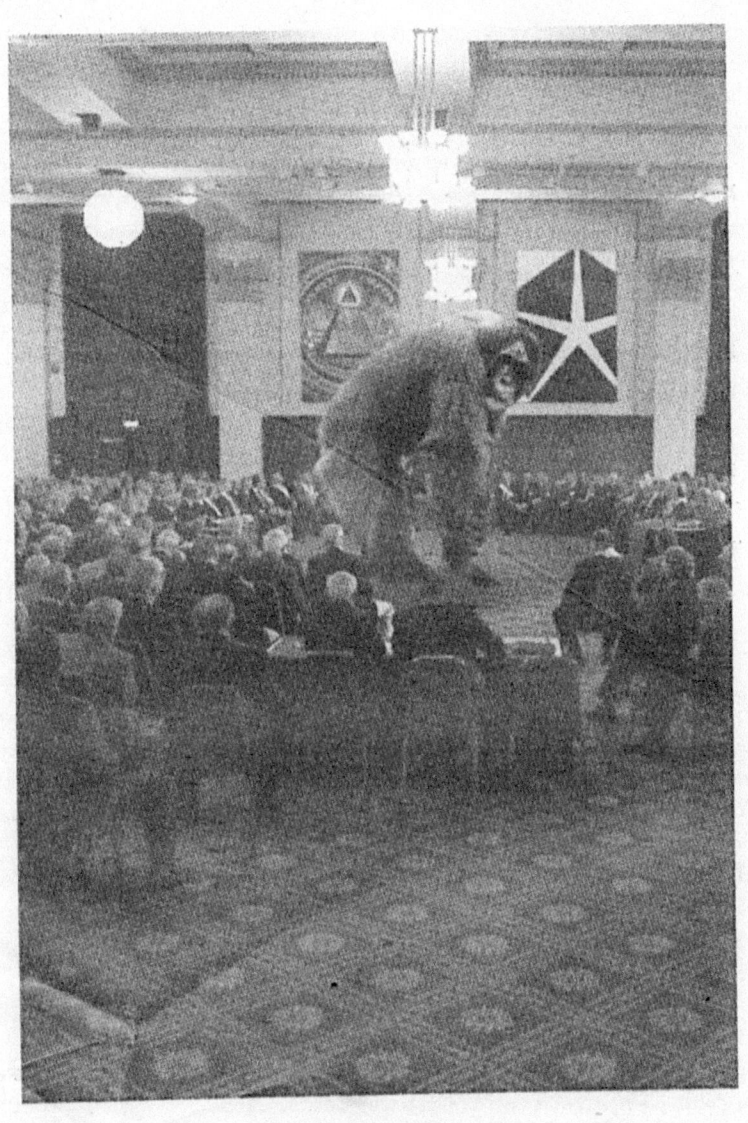

Image: Theo Michael, *The Animal in Capitalism*

all this discussion of evil and negation, Virno switches register and object in the shift from the broader discussion of human nature and evil, to a much more specific reading of certain kinds of inappropriateness as a mode for action. The bulk of the book, in fact, is made up of series of comments of that peculiar form of rule breaking known as joke making. Here Virno takes his cue from Wittgenstein's claims regarding the difficulty of rule following, as well as Freud's reflections on jokes (although, it should be noted, without accepting any of the arguments for the role of the unconscious). There is a translation difficulty here: *Witz* covers not only jokes, but wit and other forms of word play (which might explain why many of Freud's examples that Virno takes up here are not very funny). That aside, Virno takes the complicated set-up and audience of the joke to constitute a 'diagram' of 'innovative action', that is, as a model of praxis. The third party witness to the joke partakes of a kind of state of exception, and thus witnesses the creation of a new model of action. This argument, as lengthy as it is in the book, is strangely unconvincing, depending on a strangely withdrawn notion of the spectator (following Kant and Arendt) which doesn't seem vastly different from doing nothing at all, but perhaps the multitude can't do without some degree of spectacularisation after all.

Although Virno shares many of the same terms with his post-autonomist comrades – multitude, immaterial labour, exodus – he comes at them with very different sources. Not the antihumanist Deleuzo-Foucauldianism of the resolutely anti-naturalistic Hardt and Negri, but via much more classical figures – Aristotle, Schmitt, Hobbes, Wittgenstein – linking these effectively to contemporary debates in linguistics and neurology. As an intervention into current debates about capitalism, language and politics, it is relatively coherent for all that, but it suffers, as we are all supposed to (maybe), from excessive generalisation, from an over-emphasis on the generic biolinguistic traits that constitute the species (even if it is only *now* that we can realise it). Concepts of the generic have in fact been making quite a comeback of late: the return to the early Marx in Badiou's concept of 'generic humanity' and in Rancière's 'generic intelligence'. Virno has much to contribute to *this* debate with his fierce philosophical, linguistic and political armoury, but he's unlikely to bring much cheer to whatever is left of what Virno calls the post-Seattle 'global movement', even if they do, as he acknowledges, understand both 'the arena of struggle and its stake'.[6] Elsewhere he writes:

> The global movement ever since Seattle resembles a half-functioning voltaic battery: it accumulates energy without rest but does not know how and where to discharge it.[7]

This is really the heart of the problem: all this energy but very little systematic analysis. The supposed historical specificity of the possibility of making generic claims about human nature says very little about how we got to where we are, nor where we go from here. Virno is usually careful not to privilege one kind of work or worker over another:

> When I speak of a 'mass intellectuality', I am certainly not referring to biologists, artists, mathematicians, and so on, but to the human intellect in general, to the fact that it has been put to work as never before.[8]

Nevertheless, by pitching his discussion of 'innovation' and 'negation' at such a level of abstraction, Virno runs the risk of turning a diagnosis into a template, not for activists, but for those who seek to turn a profit from such an understanding of biolinguistic capitalism.

Info

Paolo Virno, *Multitude: Between Innovation and Negation*, trans. Isabella Bertoletti, James Cascaito and Andrea Casson, Los Angeles: Semiotext(e), 2008

Footnotes

[1] 'Natural-Historical Diagrams: the "New Global" Movement and the Biological Invariant', trans. Alberto Toscano, *Diacritics*, forthcoming, Winter 2008 (originally from *Forme di Vita*, 2004).
[2] Ibid.
[3] Karl Marx, 'Economic and Philosophical Manuscripts of 1844' in *Early Writings*, trans. Rodney Livingstone & Gregor Benton, London: Penguin, 1975, p.453.
[4] Steve Wright, 'Reality Check: Are We Living in an Immaterial World?' at, http://www.metamute.org/en/Reality-check-Are-We-Living-In-An-Immaterial-World
[5] Silvia Federici, 'Precarious Labour: A Feminist Viewpoint' from, http://auto_sol.tao.ca/node/3074. See also David Graeber's 'The Sadness of Post-Workerism or Art and Immaterial Labour Conference: A Sort of Review', where he makes the more forceful point that:

One could, even, start from the belated recognition of the importance of women's labor to reimagine Marxist categories in general, to recognize that what we call 'domestic' or even 'reproductive' labor, the labor of creating people and social relations, has always been the most important form of human endeavor in *any* society, and that the creation of wheat, socks, and petrochemicals always merely a means to that end, and that – what's more – most human societies have been perfectly well aware of this. One of the more peculiar features of capitalism is that it is not – that as an ideology, it encourages us to see the production of commodities as the primary business of human existence, and the mutual fashioning of human beings as somehow secondary.

From, http://www.commoner.org.uk/wp-content/uploads/2008/04/graeber_sadness.pdf
[6] Paolo Virno, 'Natural-Historical Diagrams', op. cit.
[7] 'Interview with Paolo Virno', conducted by Branden W. Joseph. From, http://info.interactivist.net/node/4982
[8] Ibid.

Nina Power <n.power@roehampton.ac.uk> is a lecturer in Philosophy at Roehampton University

MONSTROUS PLANS & GOOD HABITATS

Was modernism complicit with colonialism, and did the struggle for decolonisation also entail the targeting of imperial modernist architecture? **Mark Crinson** visits the exhibition In the Desert of Modernity to see if the charge will stick

Modernist architecture was full of good intentions. It would dispel the irrational and the merely woolly. It would cleanse the body and heal the soul. It would draw a world of nasty parochialisms towards cosmopolitanism and the international. Announcing a new beginning, it dismissed bad objects like history and style, as well as old regimes and empires, as just so much detritus of the past. And for some time, at least in Europe, it seemed possible to maintain this illusion. But then came the fall. Modernism was bureaucratised and commercialised, architects rebelled against its constraints, residents rejected its harsh disciplines, and historians began to expose its tainted connections. Yet modernist thinking retains some of its attractions; there is a seductive asceticism about it still, and nostalgia often for its utopias.

One of these lost illusions is addressed by the exhibition In the Desert of Modernity, held at the Haus der Kulturen der Welt in Berlin (29 August – 26 October 2008). It argues for connections between French colonialism in North Africa, modernist architecture and planning, and social unrest in the estates, *bidonvilles*, and *banlieues* of contemporary European cities. Essentially, the exhibition's thesis is that colonial attitudes to North Africa, as a laboratory of modernity, generated forms of resistance to colonialism and internal critique of modernism, and that all of these – the laboratory, the resistance and the critique – were then imported into European cities. Such an ambitious argument would potentially overload any exhibition, but the photographs, plans, videos, letters, posters, paintings, magazines and books on display did just enough to suggest the richness and significance of the subject.

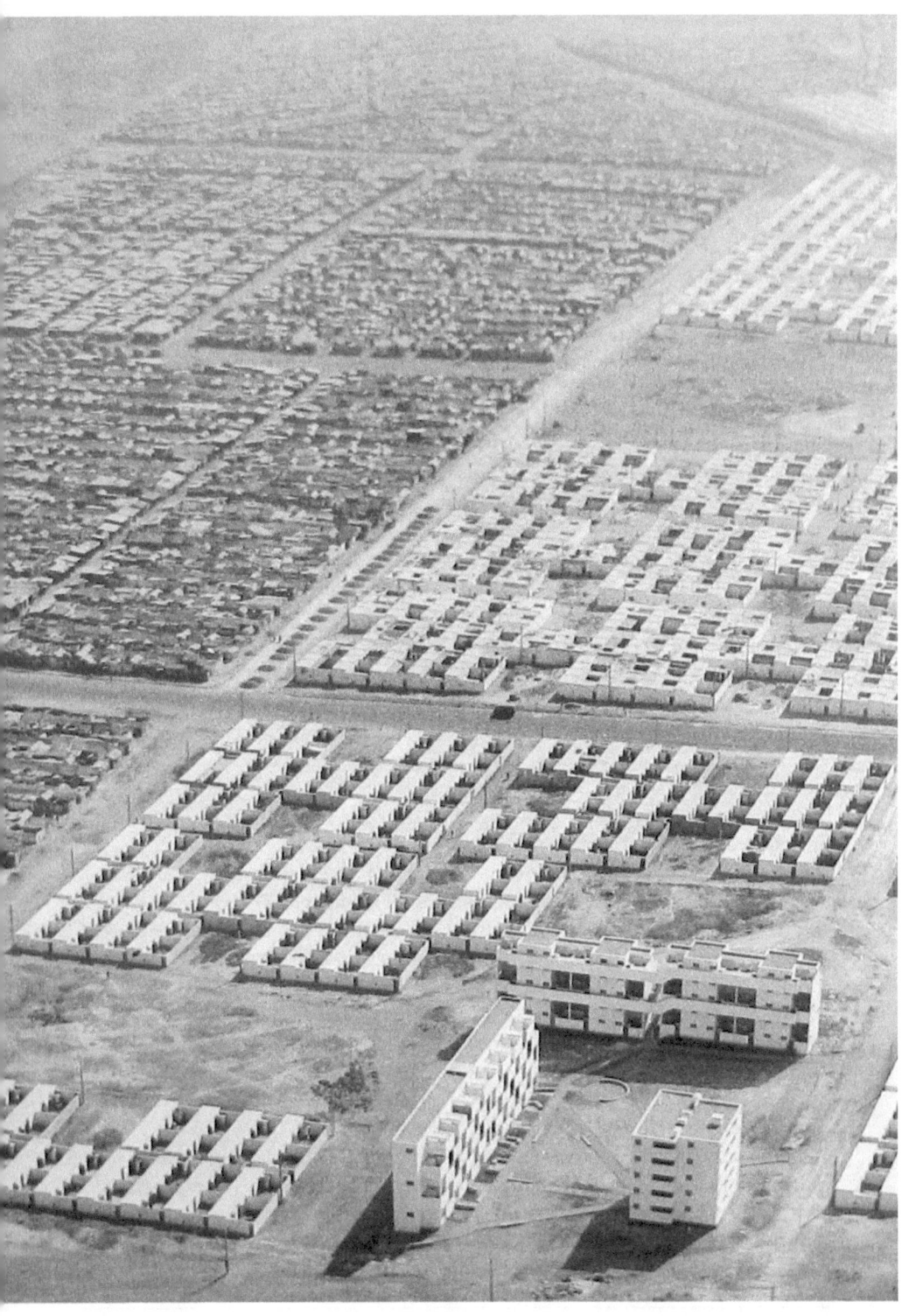

Image: Aerial view of the Carrieres Centrales, Casablanca, c. 1953, showing Cité Verticale surrounded by estates of courtyard houses and (left) the *bidonville*

Monstrous Plans & Good Habitats

Berlin might seem a relatively neutral venue for this show (certainly 'imperial Berlin' has different resonances than London's 'heart of empire'), and apart from a plan of the bewildering casbah-like layout of the Freie Universität Berlin, there is little to link Berlin with the story told here. But one 'desert' suggested by the title might be the ruined city of Berlin after the WWII, out of which rose the building that houses the exhibition. Originally called the 'Hall of Congresses', its foundation stone was laid in 1956 at the same time as that other 'desert', colonial Morocco, was enduring its final moments under colonial rule. The Hall has its own history of modernist over-reach. With its spectacular double ring beams, it was originally much-lauded for its daring and was seen as a symbol of liberty in sight of East Berlin, but the structure failed and it crashed to earth ignominiously in 1980. Rebuilt, it now houses the Haus der Kulturen der Welt, which immediately resonates more with its near-contemporary, London's Commonwealth Institute – another spectacular roofscape whose rhetoric was modernism's capacity to break free of old thinking. The evocation of this '50s ambience of late colonialism, and modernism in its last spate, is also conjured up by the exhibition's spindly stands and stencilled lettering.

There are two organisational premises in the exhibition, one taking us through and the other cutting across it. The first hops across a period from the early 1950s to the present. Following the dictum of an earlier governor-general of Morocco, Hubert Lyautey, that 'a construction site is worth a battalion', the exhibition starts with post-war Casablanca's vast mass-housing projects (known as the 'monstrous plan'). These were laid out by Le Corbusier's disciple Michel Ecochard for Casablanca's swollen population of migrant labour and located well away from the French areas. Ecochard sponsored the zoning policies of the Congrès Internationaux d'Architecture Moderne (CIAM), but he also brought in geographers and sociologists and he allowed some of his younger architects their heads. Two projects by them are focussed on here – Sidi Othman by Studer and Hentsch and the Cité Verticale by Candilis, Woods and Bodiansky, both practices with strong connections to Le Corbusier. Then the exhibition turns to the *bidonvilles* (shantytowns) of North Africa, both as subject of study by planners and architects and as centres of anti-colonial resistance. An explosion of aptly-titled books and exhibitions are seen to follow from this: the Museum of Modern Art's *Architecture Without Architects* (1964), Sibyl Moholy-Nagy's *Native Genius in Anonymous Architecture* (1957), John F. C. Turner's *Housing by People* (1977), and so on. This often romanticised view of 'spontaneous' building methods and living patterns was taken back to Europe where it served to reinvent modernism for housing, tourist resorts, and the odd university. Finally, the exhibition turns to the housing of immigrants from the old colonies and the new wave of more informal mass housing projects in the *banlieues* of French cities. Designed in 1960 by the same architects who designed the Cité Verticale, Toulouse-le-Mirail was one of these, and is shown here first with a gloriously embarrassing photograph of the architects and their Team 10 friends desporting themselves in front of it, and later as one of the flashpoints for the nationwide riots of 1998 and 2005.

The exhibition's second organising premise is a series of oppositions working to rhetorical effect through the exhibits. There is the spareness of architectural rendering and the stark mess often depicted in documentary photography. Also the informality of contemporary academics talking to camera and the stiff paternalist voice-overs of propagandist film from the time. The most blatant of these oppositions is, seemingly, the two paintings in the show, one a Le Corbusier, the other a work by the little-known Moroccan painter Chaibia Tallal. The contrast appears blatant: Le Corbusier, the male colonial oppressor, paints the bare-chested woman of orientalist dreams; Tallal, a self-taught Moroccan artist who grew up in the suburbs of Casablanca, was taken up by French museum curators and had links with the COBRA group, shows a woman swathed in bright drapery. Yet her work plays into the myths from which Le Corbusier drew; particularly North Africa as a place of direct, Matissean sensuousness.

the universalist belief of modernism switched to the culturally particular

The heart of the exhibition is undoubtedly post-war Morocco. Historians have recently been burrowing away in the archives to detail the profound shift that occurred in CIAM, beginning with the meeting in 1953 and the internal apostasy that would gel into the Team 10 group. The presentations at that meeting by two groups of architects based in North Africa – GAMMA and CIAM-Alger – brought something entirely new to modernist thinking. This was the idea that one could learn from the *bidonville*. It wasn't that these groups were particularly political in their approach – they weren't campaigning against under-building or exposing the exploitation of workers or even the separation of Moroccan living quarters well away from the French areas. Nevertheless their presentations caused uproar at the CIAM meeting. Vernacular adobe buildings in North Africa had long been admired by orientalists and modernists alike, but to treat the *bidonville* with respect was to accept that there could be a contemporary vernacular which wasn't picturesquely fixed in some sun-drenched view of the Maghrebian medina, but was instead an expression of changing relations between city and country. The *bidonvilles* were built by and for the same construction workers who were engaged on the colonial expansion of Casablanca, and they were made from the remnants of that official building work – tin, lumber, industrial waste.[1] They could be messy and they were certainly poverty-stricken. But instead of treating the *bidonvilles* as a problem or a symptom, as the French had done since they first appeared in 1907, these groups saw them as complex products of acts of building and acts of dwelling. They might even provide design solutions, but their understanding would certainly have to be multidisciplinary. So the universalist belief of modernism was switched to the culturally particular; the olympian view to the view from the street; zoned and separated spaces to in-between and clustered spaces; and veneration of the hard sciences of technology and medicine redirected to the soft sciences of anthropology and sociology. As the exhibition puts it, 'habitat' replaced the 'machine for living'.

Of the housing schemes outside Casablanca that have a prominent place in the exhibition, the Nid d'Abeille, designed by members of the GAMMA group and part of the Cité Verticale, is at first a baffling design. Its white walls seem cardboard-thin, as if a model has been montaged onto a photograph – an impression enhanced by the brightly-coloured inner walls of what seem to be balconies. There is a giant rhythmic pattern of blank rectangular panels projecting out of the surface of the building. The scale seems wrong. There's nothing to measure the size against, and if these are balconies there are no windows visible above or beyond them. It's only by looking at an aerial photograph of the wider setting showing the surrounding low rise buildings of the *bidonville* that you realise that these oddly-scaled balconies are really the walls of courtyards that aren't intended to be looked over. Each projection is therefore a courtyard suspended from the building, and the courtyards and their inward-looking apartments are stacked up. The architects envisaged that the interior spaces of the apartments could be made flexible by having movable walls, so enabling subdivision according to the size of the family or the necessity for different sizes of shared spaces.

> the idea that North African housing projects 'migrated' to European cities is striking but unsustainable

By almost any estimation the Cité Verticale should actually rank as one of the least mendacious of modernist public housing schemes, and indeed Moroccans were drawn by its symbolism and its promise. It was much more generous with space than the public housing that succeeded under Morocco's post-colonial administration. Despite its modernist abstraction, the Cité Verticale was clearly based on studies of vernacular housing and culturally-specific typologies – those high balconies give the apartments an inward-looking privacy and the visual sense, at least, of property as inalienable. However, it is difficult to see anything learnt

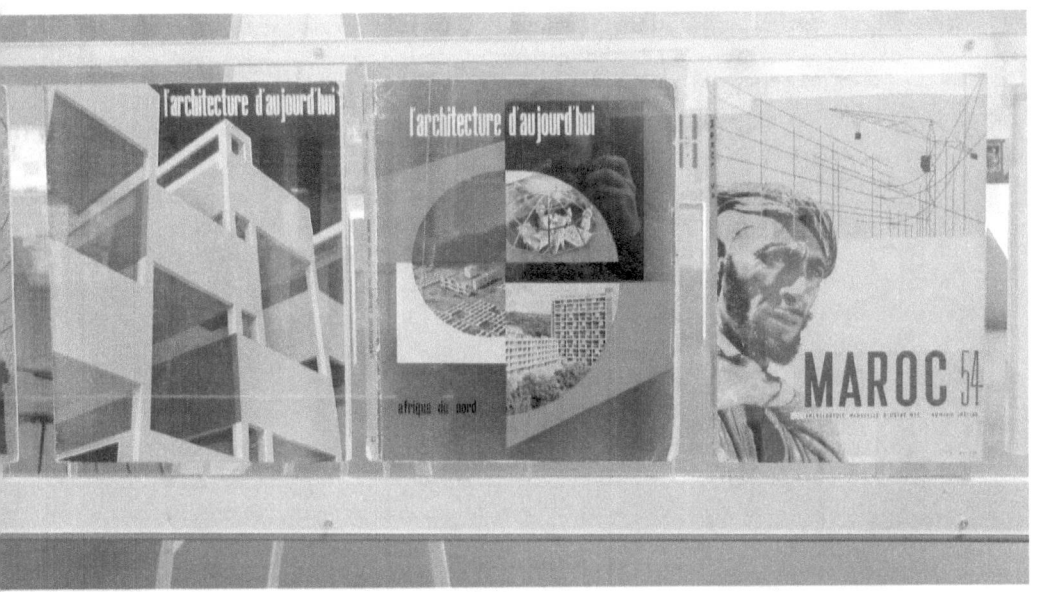

Image: Gallery display of issues of *L'Architecture d'Aujourd'hui*

here specifically from the *bidonville*, unless it is that the rudimentary provision of subdivided space allowed for appropriation by the residents. And there's a further twist to this. If residents' adaptation of a building can be understood as a kind of critique of it, then the present state of the Nid d'Abeille is striking. All the double height spaces above the patios have been bricked in and made into livingrooms or bedrooms.[2] Was this because the residents had moved beyond the need for even an abstracted vernacular, much as French theorists thought they should, and that the Nid d'Abeille was a transition into this modernity but in ways its architects didn't predict, an architecture of evolution for those the French called the *evolués* for this very reason?[3]

The problem remains that the blocks were built far from the city and far from any contact with the French colonists. And the valid question to ask here – one that the exhibition does not broach – is whether this really has anything to do with modernism. Certainly, there had been plenty of non-modernist housing schemes which were also built under these conditions of virtual urban apartheid – mimetic neo-Moroccan schemes, for instance, concordant both with orientalist views of the serendipity of Muslim life and with colonial policy, learnt from the British, of a dual cultural system.[4] In relation to this, one of the interesting insights that the drawings show is that some of the unbuilt modernist schemes from the 1950s had slightly different treatments according to whether they were intended for Muslims, Jews or Europeans (Spanish and Portuguese labourers also lived in the *bidonvilles*, and their existence explains the 'European' variants on these designs).[5] Each design shows a certain crude representation of different degrees of openness: Europeans have more windows and therefore are more accepting of private-public merging, Muslims look inwards but need shaded outdoor space, and Jews would have a latticed courtyard because presumably they are somewhere between the two. A 'mixed' treatment was even supplied for families of different backgrounds that would combine all the previous three variants.[6] This wasn't a taxonomy by race, caste, culture or religion, or at least not by any one of those categories to the exclusion of the others. In some ways it was closer to pre-modernist colonial taxonomies that were often multiply coded (such as in the layout of the housing for New Delhi). It could be understood as an early manifestation of multiculturalism; not the hoped-for universalism of modernism, but a kind of blurred relativism.

As the preponderance of French phraseology shows, this is basically a Francophone story. The exhibition's broader gestures and claims of wider influence are often unconvincing, and sometimes it seems that there are simply too many things piled up on top of one another and not enough real relationship between them. The idea that housing projects designed for North African cities 'migrated' to the outskirts of many European cities, for instance, is immediately striking for its reversal of the old model that every good idea was exported to the colonies – but on any closer examination it is unsustainable.

Rather than accept the exhibition's gesturing towards a European-wide impact, suppose one were to challenge part of the thesis in the case of a country other than France. How far was modernism complicit with colonialism in, say, Britain? How much was it part of a pervasive colonial way of seeing? It is striking how many of Britain's post-war planners, for instance, were brought up in the British colonies or how many British architectural practices still had colonial commissions on their books well into the 1960s. There are resonances, to take another example, between British welfarist housing and anti-insurrectionist and anti-communalist re-housing policy in colonial Malaya during the so-called 'Emergency'.[7] And the French Protectorate's policy of deliberate under-building for its Moroccan workers was certainly shared by British colonialists (it can be found in New Delhi as well as in the Persian Gulf company towns of the Anglo-Iranian Oil Company – present-day BP). The work of Alison and Peter Smithson might be another place to look at this, especially as their 1953 CIAM Grid is represented in the exhibition. Though anti-modernist, or least

Mark Crinson

modernist apostates, the Smithsons never seemed to have expressed any particularly anti-colonial views. To some extent their interest in the everyday life of working class areas like London's East End was based on techniques of colonial anthropology (via the Mass Observation project). And they certainly became enthusiastic advocates in Britain of the North African modernists' architecture and their studies of the vernacular. To get from here to the much later populating of the Smithson's housing estate at Robin Hood Gardens, in part by migrants to Britain, is a rich and complicated story, worthy of an exhibition in itself. But there is certainly no cause and effect relation between being influenced by colonial ways of thinking and having architecture, several decades later, lived in by people, some of whose families may have come from ex-colonial countries. More relevantly, if Robin Hood Gardens can be said to stand for anything in this context it is that the difference between high modernists of the CIAM sort and modernist apostates of the Smithsons-Team 10 sort was simply not enough of a difference; cultural relativism was only relative. Anyone unversed in architectural history would be unlikely to see any significant difference, other than small formal variations, between Robin Hood Gardens and other mass housing estates of the '50s and '60s. The borders and protocols of what an architect does and of what modernist architecture was about, and how both work within the modern state – welfarist, colonial, neoliberal, neo-colonial – were never at stake in 1953 and hardly at all since.

there was not enough difference between high modernists and modernist apostates

Finally, the exhibition's images of street demonstrations need some comment. The idea that anti-colonial consciousness was bred mostly in the *bidonvilles* of Casablanca fits the thesis rather too neatly; in fact Salé and Fez, both 'traditional' cities, were far stronger nationalist centres.[8] But there are wider problems with only thinking of resistance in this way. The mirroring of images of demonstrations in Morocco in the mid-1950s and in France 50 years later first suggests important links between them as anti-colonial struggles, and also sets up the all-too-easy idea that the architectural environments are much the same and that the environment is an essential part of the cause. Neither statement, it would seem, needs arguing through. In this regard we must remember that the actual amount of modernist housing in Morocco by 1956 was tiny and that infinitely more significant was the policy of urban apartheid. A more general point is that these images of public, overt acts of resistance in the exhibition appear to show that they only occupy the environment incidentally rather than as active parts of it – the townscape is mere backdrop to their drama. And lastly, despite gesturing towards it, the

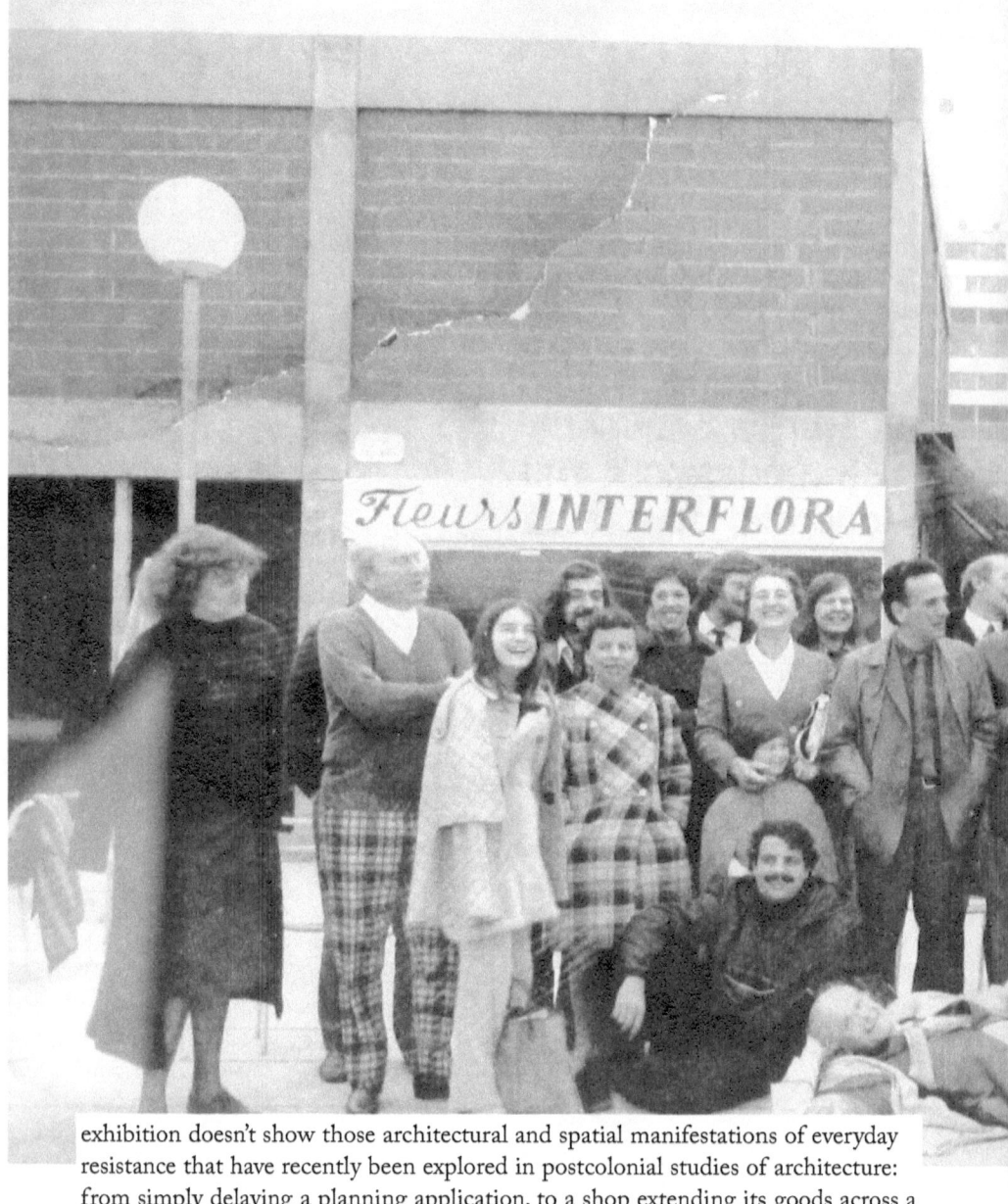

exhibition doesn't show those architectural and spatial manifestations of everyday resistance that have recently been explored in postcolonial studies of architecture: from simply delaying a planning application, to a shop extending its goods across a pavement, nannies treating a public square as a meeting place, or residents keeping chickens on a balcony.[9] Such modest actions, one might suggest, create far more everyday disruption in their constant friction against the ordinances of a dominating culture than any situationist *dérive*, and are at least as much about the carving out of counter spaces of living as any street rally.

Image: Cité Verticale Team 10 members on a visit to Toulouse-le-Mirail, 1971

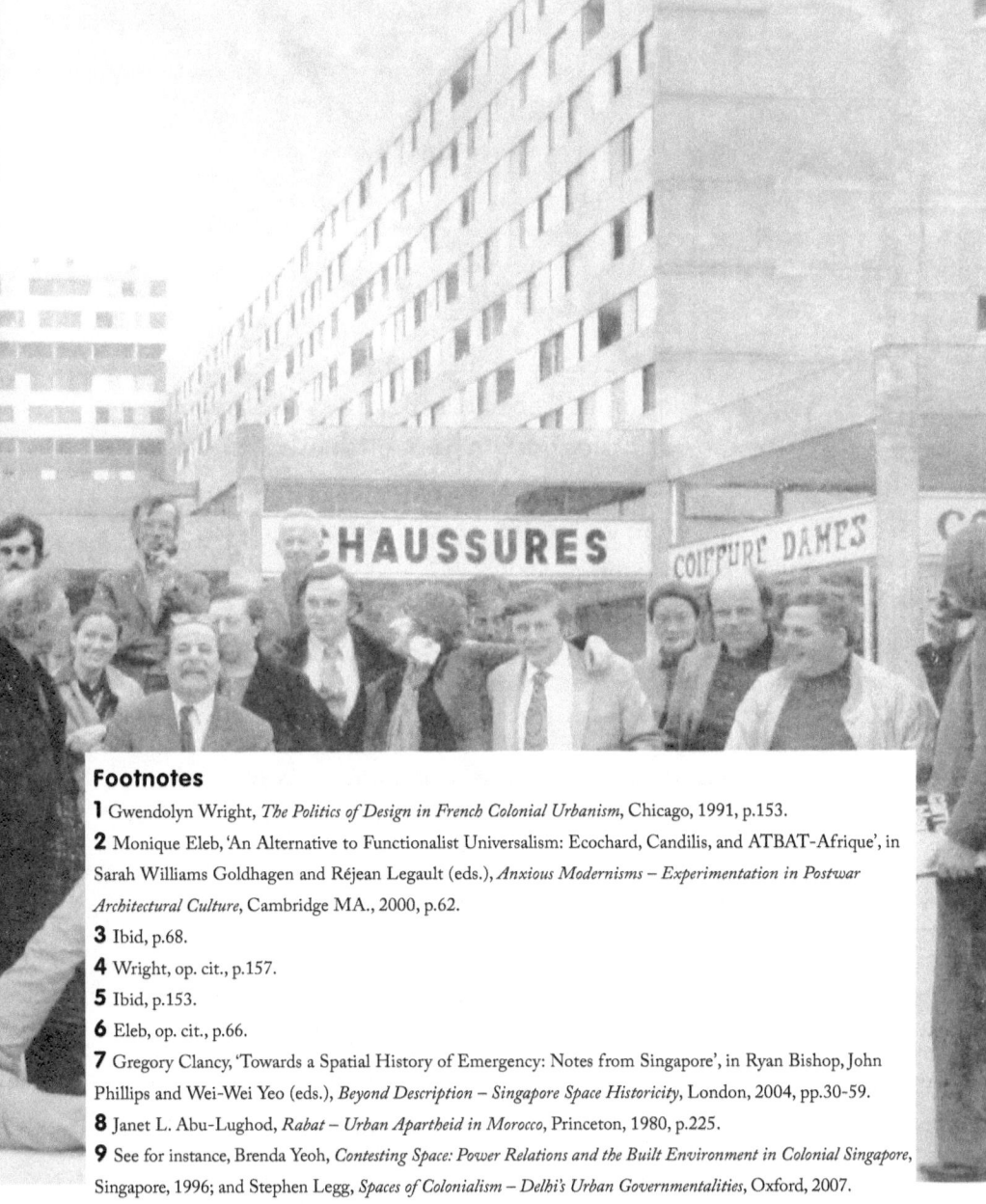

Footnotes

1 Gwendolyn Wright, *The Politics of Design in French Colonial Urbanism*, Chicago, 1991, p.153.

2 Monique Eleb, 'An Alternative to Functionalist Universalism: Ecochard, Candilis, and ATBAT-Afrique', in Sarah Williams Goldhagen and Réjean Legault (eds.), *Anxious Modernisms – Experimentation in Postwar Architectural Culture*, Cambridge MA., 2000, p.62.

3 Ibid, p.68.

4 Wright, op. cit., p.157.

5 Ibid, p.153.

6 Eleb, op. cit., p.66.

7 Gregory Clancy, 'Towards a Spatial History of Emergency: Notes from Singapore', in Ryan Bishop, John Phillips and Wei-Wei Yeo (eds.), *Beyond Description – Singapore Space Historicity*, London, 2004, pp.30-59.

8 Janet L. Abu-Lughod, *Rabat – Urban Apartheid in Morocco*, Princeton, 1980, p.225.

9 See for instance, Brenda Yeoh, *Contesting Space: Power Relations and the Built Environment in Colonial Singapore*, Singapore, 1996; and Stephen Legg, *Spaces of Colonialism – Delhi's Urban Governmentalities*, Oxford, 2007.

Info

In the Desert of Modernity, Haus der Kulturen der Welt, Berlin, 29 August – 26 October 2008

Mark Crinson <mark.w.crinson@manchester.ac.uk> lectures at the University of Manchester. He is the author of *Modern Architecture and The End of Empire* (Ashgate, 2005)

THE POLITICAL IMMUNITY OF DISCOURSE

The English translation of Roberto Esposito's *Bios* appears to be an important contribution to the critical analysis of a politics of life, but can the book's claim to 'revitalise' politics really be thought from within the exclusive bounds of academic philosophy? **Erik Empson** reviews

I dare say it is not me alone who, when discussing the nature of biopolitics with the uninitiated, has found himself rather meekly repeating formulas such as 'politics become the management of life itself' or 'control over the body has come to be the object of power', whilst feeling that they fall short of capturing the specificity of what more subtle commentators have discussed at length – the change from disciplinary societies to those based on control – as a genealogy of modern industrial societies. If modern social and political life can be rightly called biopolitical, we need to find an apposite definition of what this means on its own terms, without having to first explain the history of what came before and what is so different about today.

A sympathetic view of Roberto Esposito's book, *Bios: Biopolitics and Philosophy* would be that he aims to do just this: produce an affirmative form of biopolitics based on a nuanced philosophical understanding of both sides of the equation 'life' and 'politics'. Arguably philosophy is more readily suited to this task than any other discipline, or perhaps we've exhausted other approaches to the degree that this is the only one we have left. But one wonders if it can address the particular conundrum of biopolitics that Esposito sets out to solve, namely 'why does a politics of life always risk being reversed into a work of death?'

Whilst much recent critical scholarship has tried to detail the untold history of this transition, Esposito's question presupposes it. It takes for granted that there is something universal within the biopolitical that can be drawn out and used to qualify and moderate ethical action and put it to service for the political health of a community. Esposito's initial survey is broad; it features discussions about a child's right to sue for not being aborted in France in 2000 and the 'humanitarian'

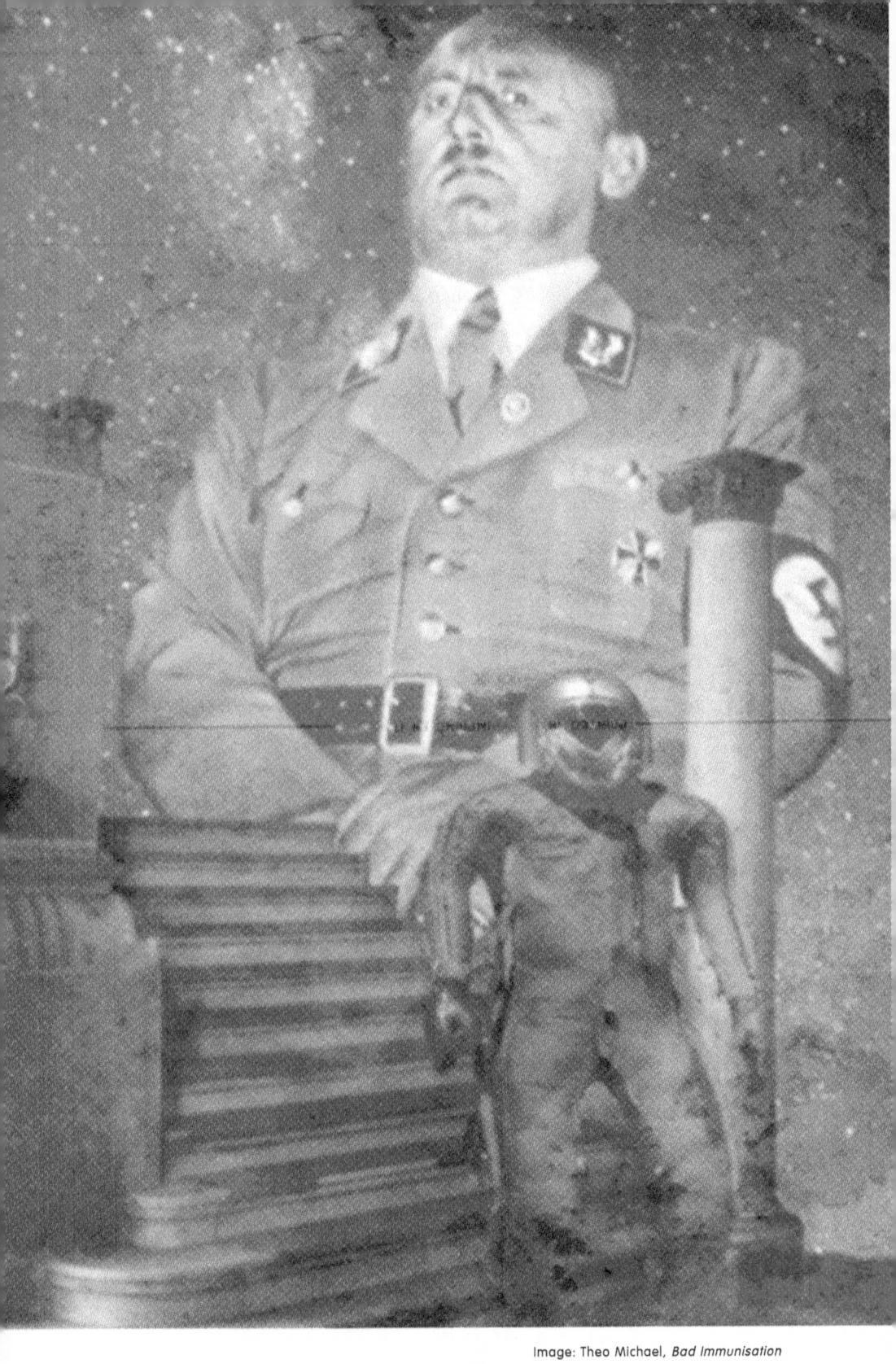

Image: Theo Michael, *Bad Immunisation*

war in Afghanistan in 2001, the police massacre of Chechen hostages in Russia in 2002, rape in Rwanda in 2004 and China's single child policy. These are the type of parables that dominate bioethics, innervating and encouraging the participation of an otherwise apathetic televised public in the spectacular interplay of complicated policy decisions: the deceitful managerial languages that disguise expediency behind masks of human concerns.

The idea that government directly entails a politics of, or over life, is fast in danger of becoming a platitude (of more interest is that political science, business studies and cultural criticism have internalised this discourse reflexively by understanding their project in such terms). Like Esposito, we begin to ask the question: 'wasn't it always?' The fact of biopolitics doesn't seem to help us to get to the truth of biopolitics. Maybe biopower is only ever local, historically circumscribed, situational, bound to the technologies available, a particular method employed to differentiate who can and who cannot live. Given the existing genealogies of power here, and remembering for a moment all those that didn't make it onto Fox News, what can philosophy add to an understanding of the process on a generic level (if such a level does in fact exist)?

why does a politics of life always risk being reversed into a work of death?

The guarded back-cover endorsement of Michael Hardt and Antonio Negri notwithstanding, this is not another book that will take up biopolitics through the increasingly familiar lenses of changes in subjectivity and class composition. In fact Esposito's object is both narrower and wider. He wants to revise political theory and make that revision central to a new analysis of modernity: 'contemporary thought cannot fool itself in belatedly defending modern political categories that have been shaken and overturned.' Promising enough, unless it's just another thinking away of difficult political realities like class or inequality.

The particular category Esposito has in mind is sovereignty. But in claiming that biopolitics is the antithesis to sovereignty (or at times the substitution of it) and thus the linchpin of the modern age, Esposito is only reiterating what others have said before. If, as his language suggests, he is speaking to the converted, the question is of the 'do bears shit in the woods?' variety. However, in claiming as he does that *immunisation* is the true motor of the transition to a biopolitical paradigm, he is saying something distinctive and new. Although whether it is useful, real or true is another matter entirely.

By highlighting and elevating one particular 'logic' out of the fairly abundant selection available to us to explain modernity, and especially because it concerns a

Erik Empson

pretty bloody serious issue – the emergence of strategies for the treatment of 'life unworthy of life' (the Nazi phrase Esposito uses consciously and deliberately) – it seems inexcusable that the author fails to mark early on the particular sense of immunity he is employing. Immunity can mean exemption from liability in a legal sense, or the biological process where the presence of antibodies prevents a disease. While we may have optimistically hoped that the author would bring these two meanings together in a productive combination, right from the start he plays with their semantic ambivalences.

Semantic Orbits

The clean white, transparent, mechanical Machiavellian magic of the sovereign state paradigm has mutated. In the murky depths of modern biotechnological societies of control, the obscure dark arts of legalistic mumbo-jumbo, the speaking in tongues, the high and low brow geomancy now reigns supreme. What makes immunity a more useful instrument in understanding this transition than any other astrological linking of points to make constellations that become signs?

The author charts three major moments in the theorisation of biopolitics: German vitalist thought about politics, race and nation at the beginning of the 20th century, French thought in the 1960s, and the late 1970s in the Anglo-Saxon world when research institutes concerned with biology and politics, still in existence today, were established. In doing so Esposito rightly shows some of the deeper connections between the experience of fascism, the science of the state and neoliberalism's resurrection of 'natural' limits to social and political engineering. To our relief his earlier jarring reference to Nazi eugenics finds its place. These intellectual currents give over to biology what many readers of this book, should they chose to read its passages rather than the Foucault lectures of 1976 that it paraphrases, would no doubt want to claim for politics: the power to change our circumstances, our capabilities and to decide, outside the grip of 'nature', our own moral code.

This brief genealogy is fine as it goes, although there is a tendency here to only trace the appearance of the term in explicit connections between biology and politics. In fact 'immunisation' is part and parcel of the biological roots of normal, healthy social science and the biopolitical is much more than that which explicitly goes under its name. That such a large part of the origins of modern social scientific currents, progressive and reactionary alike, modelled themselves on the biological sciences means that a more systematic, comprehensive study is needed to complement the sketchy history that Esposito offers. For sure, the reason he does not do this is because his object is not the general history of the reduction of social life to being the poor relative of biology, but to show that the attempt to engineer society along these lines is intrinsic to the nature of modernity. Nazism was not a unique and abortive experiment with organising life on the basis of biological determinants. Immunisation is more than an organic metaphor, it is an expression of an increasingly general tendency to think and act politics according to biological principles.

Esposito is interested in exploring the moral codes that govern action as a dialectic between normativisation as the self preservation of a community forming a *bios*, and individuation as a mode of a community's immunisation of itself – or equally, the defence against the very norms that bind communities. This is the central point of his book but it is precisely here, on the book's apparent strengths, that its failure to clearly demarcate its terms – biological defence or legal impunity – makes itself apparent. On the one hand, to be immune is to have non-being, i.e. nothing in common with the community. On the other hand immunity is the negative introjection of 'the negative modality of its opposite' as an internal defensive apparatus of the community. Esposito likes these semantic ambivalences, this 'surplus of sense', because in the multiple derivations of sense he can have his proverbial cake and eat it; the whole can constitute itself through its parts, whilst any unpalatable slices can be left on the table.

> I do not intend to argue that modernity might be interpretable only through an immunitary paradigm [...] nor deny the heuristic productivity of more consolidated exegetical models of use such as 'rationalization', 'secularization' or 'legitimation'. But it seems to me that all three can gain from a contamination with an explicative category, which is at the same time more complex and more profound. One that constitutes its underlying premises.

The paragraph continues:

> The immunisation paradigm instead refers us to a semantic horizon that itself contains plural meanings – for instance, precisely that of *munus*. Investing a series of lexical areas of different provenance and destination, the dispositif of its neutralisation will prove to be furnished by equal internal articulations as is testified even today by the polyvalence that the term of immunity still maintains. But this horizontal richness doesn't exhaust the hermeneutic potential of the category. (p.51)

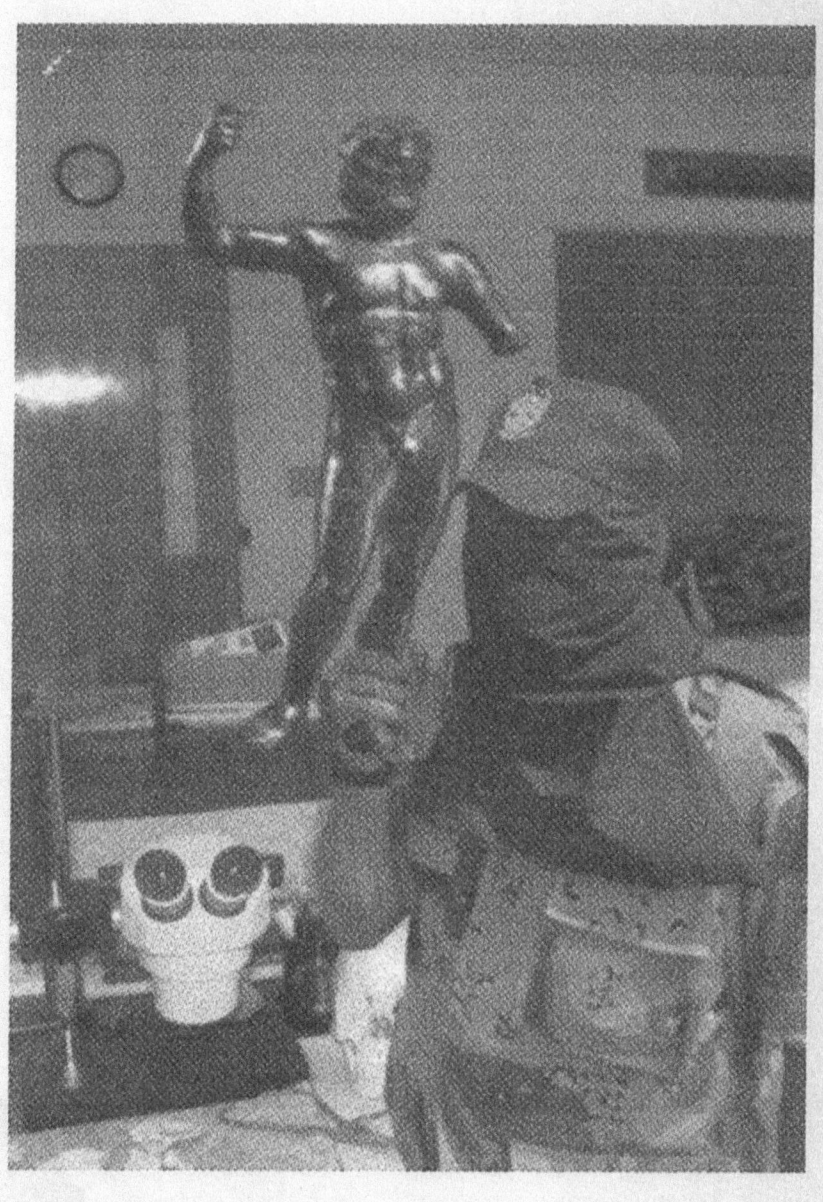

Image: Theo Michael, *The Trouble with Biopolitics*

Whereas the point of the book is to make a claim for the centrality of immunisation, its conceptual elaboration seeks to marginalise it. Esposito appears to be giving with one hand and taking away with the other. We were promised the overturning of categories but the play with multiple meanings and the obscurity of the text is beginning to suggest this book is an inconsequential verbal stunt in a semantic universe without gravity. That he should go on, as we will, to discuss the horror of the Nazi experience and its exemplary status in the biopolitical nexus, might take the fun out of the acrobatics, but there is hope.

The Death Machine

Other theorists have placed Nazism as intrinsic to modernity. One thinks of the Bauman's description of the Holocaust as the outcome of the process of technological rationalisation. What makes Esposito's contribution distinct is that by identifying three elements of Hitlerism – the absolute normativisation of life, the double enclosure of the body and the suppression of birth, in fact the least explicit political elements of it – he aims to find an exemplary instance of how important the immunisation of a community is to its apparent survival, as well as the inherent dangers of the process. Thereby he aims to isolate both the shortcomings of Foucault's analysis of the biopolitical and identify the 'approximate and provisional contours of an affirmative biopolitics that is capable of overturning the Nazi politics of death in a politics that is no longer over life but of life.'

If Esposito's challenge is to find a positive biopolitics of life, it is made all the more heavy given that he is determined to discover it in the dark recesses of a history that by and large the rest of humanity wants to forget. Either Esposito is one of the brave souls who, with all the risks it entails, wants to immerse himself in the historical problem of Nazism, not so much that it might not be forgotten, merely bearing witness to atrocity, but lest we fail to learn a specific lesson from it. Or he is carelessly and clumsily building a half-baked theory on the back of an event with enough gravity of its own to carry it off. One of the major questions Esposito poses for an affirmative biopolitics is, if the resistance of life is stronger than power, 'how do we account for the outcome obtained in modernity of the mass production of death?' Clearly his affirmation of life is not going to be totally *bonvivant*, so why do this?

According to Esposito, it is not in spite of Nazism that I have grown up (as part of the generation born in the mid-1970s) with a particular sense of right over my body, as having inherited a certain political leverage in defence of life, or living in a society where it is more or less taken for granted that life be defended. It was precisely those values that underpinned Nazi policies of eugenics and mass

Erik Empson

is this book an inconsequential verbal stunt in a semantic universe without gravity?

extermination. Indeed for the author, Nazism is the threshold of the contemporary age. Esposito makes this clear. Carl Schmitt is wrong; the specificity of the fascist state is not the making normal of the exception to the law in politics, but accepting the normalisation of biology as the object of politics: the management of the health of the nation. The showerheads that emitted noxious gases were hygienising and sanitising against the disease that had infected the purity of the German people. Authorisation for each act of ethnic cleansing came from the medical establishment, not mere pen-pushers but individuals fulfilling their Hippocratic Oath to cure their patient (the German people).

Nazism as a form of death, the indifference over life or the making of life into a machine of death is an all too easy point of reference if you are looking to find an affirmation of the human by contrasting it to its other. But it works polemically not only because it produces righteous indignation, but because the reader is existentially driven to want to find life or, one might say, hope there. Paradoxically Esposito denies us this.

As a political point of orientation the Holocaust is rightly exhausted and as a theoretical foundation, it is only the banal other of an entity, life, which (as Esposito otherwise rightly notes) is maximally differentiated within itself and thus a very different object to that established when contrasted with its other. Nazism is not the other of humanity. Its practice of death is not the other of life, but the extension of a practice of creating *bios* – a biopolitical community – in an appalling, deluded, disastrous way. For its practitioners was this wickedness not a dimension of life, a perverse joy or one of the perversions of joy?

So, immersed in the challenge of outdoing Foucault at home, Esposito arguably uncritically adopts his tendency, often latent in its semantic drifts, to speak of power in all too self-identical terms, as a subject that exists in its own right and can take decisions (p.38). And this is probably due to the fact that Esposito is concerned with the discourses that justify the state rather than the fact of the state. Indeed the clinical character of Nazi propaganda can always be contrasted with the grim, intoxicated reality of the execution of its principles of cleansing.

The analysis of 'scientific' discourse is only one side of the true picture of political life and all too often can fail to register alternative currents either suppressed by the mainstream or conve-

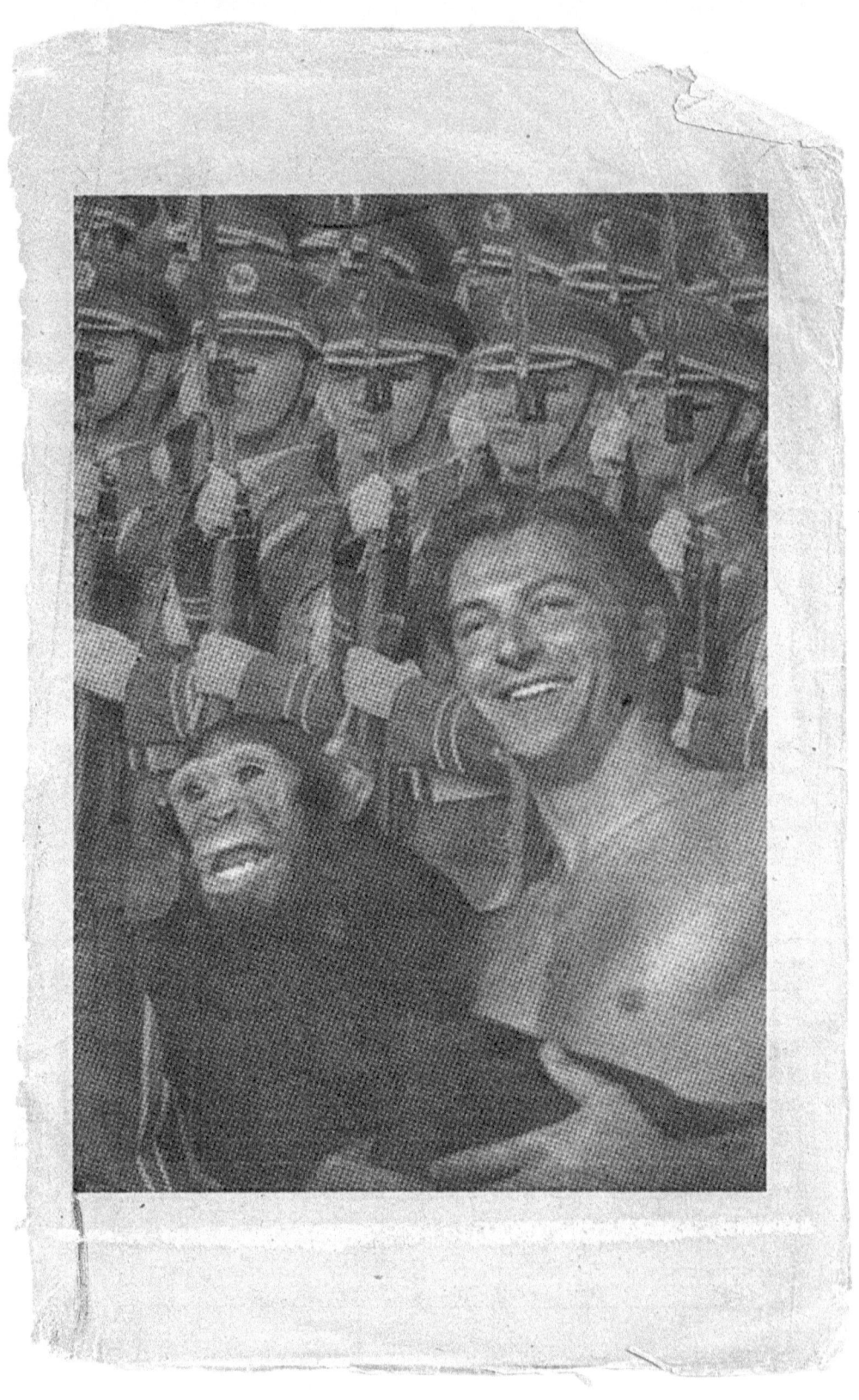

Image: Theo Michael, *Single Child Policy*

niently left out if they don't fit into the schema. And though it may be true of the Nazi State that it distributed the sovereign right to kill, it tells us nothing of the remorse felt by those that did, or the protests of those that didn't. This clearly isn't the point, but perhaps it should have been. For Foucault, power is a set of practices, institutions and languages, and resistance is also power, not opposed to it. In *Bios* we don't hear about how the immunisation paradigm is resisted. Only those practices which are within its dialectical orbit are recorded, not the refusals, the suicides and the other lines of flight that escape it. Thus strangely, although that is not his intention, we are left with a political theory that documents only *potestas*, not *potentia*, i.e. only power from above, formal constitutionally sanctioned power. Or perhaps the community-immunity nexus is so embracing that it encompasses all forms of activity. At any rate the deficit is all the more striking given that the author aims to 'vitalise' politics. We wonder which politics? Let's hope it's the good guys.

Immunising Discourse

Where Esposito revisits the canon of political theory, the acuteness of the immunisation paradigm begins to slip by the wayside. What was meant to be a genealogy becomes a clunky narrative wherein the subtlety of the historical is lost in the rhetorical. Passing through Plato, Hobbes, with a nod at Hegel, Locke (concerning man's possession of his body), Bentham and so on, the immunisation concept appears at times decidedly forced, grafted onto the subject matter. Of course, he finds the most salient expressions of it in Heidegger (whose philosophy the author is at pains to describe as the antithesis of Nazi thinking about biopolitics) and Nietzsche, who begins the special thinking about *bios*; the weak as a limitation on the strong that Esposito seeks to highlight. Make no mistake, if desperately convoluted to the point of being overbearing, if turgid to bursting point, this is still a very rich, inventive analysis. But its inventiveness borders on fiction, so we should ask what his book communicates and to whom.

Esposito's generation has a particular psychological relationship to the fascist experience; often ancillary to that is an obsession with community as the locus of social life. What are these communities that are the uncritically accepted subject of political reference? To his credit, like few others, Esposito understands that communities form on the basis of exclusion but he wants to find a way of saving the possibility of community from this fact of community. The tentative conclusion is that communities gain their strength by allowing for the proliferation of different forms of life. Ultimately the Nazi State's immunitary logic was turned against itself: protected, inoculated, the population became weak and subject to an inner and irreversible degeneration, its policy of purification for life resulting in a thanatopolitics, a regime of death. Presumably if we enact the opposite, open our communities up to the other, go beyond normativity by defending the immanence of singularity, we have an affirmative biopolitics.

Esposito expresses the political horizons of a particular generation when he claims that we can 'rise above Nazism only by knowing its drifts and precipices'. Well, will it ever be acceptable for me to say 'Treblinka' in the same sentence as 'playstation' without it being the most obscene of verbal acts? That the two are genetically connected, i.e. the latter games console being my generation's prize for not being communally integrated into society-wide clinical racial pogroms, is also an obscenity. The post-war narrowing of the *polis*, the opening up of the social, are part and parcel of the banalisation of life and the muting of politics. And if the Holocaust discourse has had a role in this it has been to modulate political extremism. The problem with Esposito is that he can understand this process in the third person but not the part he has to play in it.

The problem is not so much that biopolitics adopts a paradigm of immunisation, of creating the healthy community by delimiting its interaction and integration with others, but that biopower as a socially constitutive process of the state, adopts that project by a more thorough immersion within the paradigm: limited immigration is healthy because it adds to diversity of life practices; the multi-cultural hypostatisation of difference is the antidote to commonalities constituted on ethnic lines; moderation and the legal regulation of ethnic identity safeguards against extreme pathologies, and so on.

The difference between Esposito on the one hand and Giorgio Agamben and Antonio Negri on the other – irrespective of the former's negative definition of biopolitics and the latter's positive one – is that both Agamben and Negri are part of a broader movement for political change that is already self-reflexively embedded within the biopolitical and the questioning of the state. Esposito, on the contrary, seeks to import vitality into politics from the outside, and given that he appears to know no politics other than the prevailing or official discourse that sanctions the formation of community, it is hard to see his subject of change as anything other than the enlightened liberal public individual or the state, both suppositions of exactly that paradigm of the large-scale social engineering of *bios* that conflicts with the fact of life that each singularity is striving to exercise. It is difficult to rise to a defence of the political (against it being 'flattened into biology') that does not formalise exactly those exclusionary *dispositifs* that Esposito abjures.

The most obvious question to ask when confronted with a new discourse is: who is its subject, who is its object? Why does Esposito want to vitalise politics and what would that mean? Does he want to immunise society against fascism? Could it be that his intentions are not altogether as sincere as we have made out?

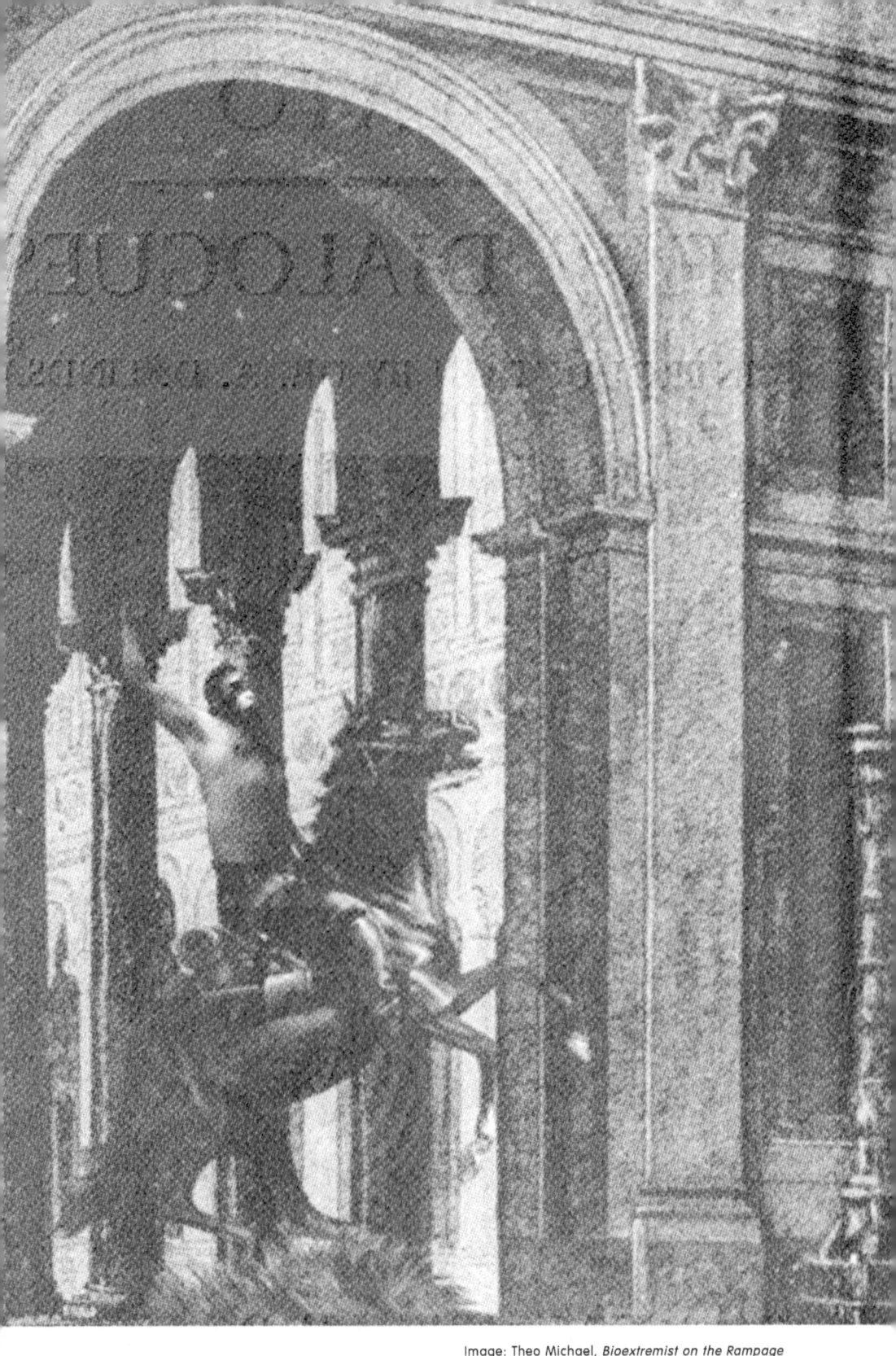

Image: Theo Michael, *Bioextremist on the Rampage*

To turn the tables somewhat, it could be argued that the most salient example of the paradigm of immunisation in relation to modernity is within discourse itself. Having dug about amongst the swollen, stinking corpses of various biocratic intellectual contributions to statecraft, Esposito knows better than most the complicity between developing ways of thinking and developing ways of acting. So it is striking that there is such an exclusionary tone to his discourse. And although my argument runs a risk of being seen as anti-intellectualism, I shall pursue it.

For all of its messing about in the sandpit of polyvalence, Esposito's discourse is not a coherently pluralist one, but peculiarly an objectivist one that inoculates and defends its personal identity through the deconstruction of its own positioning. 'To avoid death, life must be reborn in differentiated, individuated ways.' Quite so. And to avoid academic death, theory must be reborn in differentiated, individuated ways. Any other metaphor of immunisation and inoculation selectively taken from man's inglorious and homicidal past of purity and extermination can also be readily applied to high brow academic philosophy. And to stretch our criticism as far as it goes, in order to inoculate a treatise, it needs to find general defenses against particular attacks, and it does that by injecting itself with particular antibodies that have something of a common currency. With Esposito's work, the reader's expectations of what a nuanced treatment of life over death might look like in positive terms are so unrewarded we are actually thankful that he chooses the safe but pusillanimous option of finding the solution in Spinoza. Safe because that latter has been appointed by post-autonomia as the true father of anti-modernity; pusillanimous because this could have been an intellectual point of departure rather than arrival.

This is not a book for the masses, the like of which we desperately need, but one targeted at an epistemic community that excludes them. It often slips into a language that appears to suggest that it is by correcting discourse that we somehow resolve the problems of the world. Indeed his translator arguably makes the same error when he claims that the current political crisis is 'the result of a collective failure to interrogate the immunitary logic associated with modern political thought'. Crikey! The success of humankind's defence of life will depend on who reads and understands Esposito's book? Well surely if it had such a role, it should have been written in a form that had a real bearing on its content. For a book that claims to vitalise politics, no matter how suspect that sounds, to be convincing its style must be equal to its theoretical task – and with this subject it would entail a poetry, not clinical shorthand. Books need souls. It need not have more popular appeal, but it needs the 'prose of thought' to be more than a series of disjointed reinterpretations of existing fact and fancy.

Actually for all of its post-structuralist rhetorical flourishes, *Bios* reads like Hegelian ruse, an attempt to conjure out of the black hat of the negative the white rabbit of modern sensibility. Yet it lacks the inner coherence of the dialectic. For the most part it is dimly aware of its own point of arrival – or worse, frightened by what it finds there – and in the breaking of this circle, the loss of the spontaneous simultaneity of the phenomenological and transcendental (the poetry of the text), it founders in its own uncertainties. And the dialectical inversions turn into pointless paradoxes, mediocre and defensive. It poses a micro-meta-narrative but allows itself to remain in pieces and fails to labour to unify speculatively what it does rhetorically.

So how did a philosophical book about life turn out to be so lifeless? One factor is the obsolescence of philosophy itself unless it can somehow practically meld critique with insurgency. Rather than do this, *Bios* is just one in a line of roundabout affirmations of the same conclusion about the discipline: its irrelevance. Esposito's problem is a self-negating one: it is impossible to affirm life in the abstract but only ever as many forms of actual life. Subject and object are annihilated in one fell swoop and because the political is thus rendered so infirm it is paradoxically only by a return to natural right that its role can apparently be resurrected.

Philosophy can only deal with the question of life if society guarantees life in general and the position of the philosopher. The distance needed for life to become an object of philosophy is one that refuses the possibility of the question of life being a practical one, its criticality is affirmed not by the word but by the social conditions in which the word is expressed. If life is directly contested, through civil war or ethnic cleansing, or through western intervention or through lack of western intervention, you do not need philosophy but religion, not deconstruction but ideology. In the absence of practical philosophical alternatives, these are the solutions people are turning to *en masse*. This kind of academic philosophy is a peculiar kind of indulgence, and one, to allude to Marx's infamous remark, probably best done in private.

Info
Roberto Esposito, *Bios: Biopolitics and Philosophy*, University of Minnesota Press, 2008

Erik Empson <erikempson@hotmail.com> holds a DPhil in Social and Political Thought from the University of Sussex and is one of the founders of Generation-online.org. As well as writing and editing for a living he teaches part-time at the University of London

> Capital needs to sustain the fantasy of its health, efficiency and inevitability at all costs. As the crisis broadsides this fantasy, the spin-doctors are scrambling to reconstruct it. Now is our chance to stop them, writes John Barker

WISHFUL THINKERS OF THE CALAMITY BAZAAR

These really are unprecedented times, even for folk who – amateurs that we are – could see that the levels of debt which were integral to what has been called 'free market' capitalism were unsustainable. The sheer scale of the figures involved become ever harder to grasp, and were not predicted by most amateurs or, as it happens, the full gamut of finance professionals who, it turns out, appear to have believed in their fantasy world. The assessment of losses has increased from $400 billion at the end of February '08 (US Monetary Policy Forum), to $2.8 trillion as given by the Bank of England in November, via $1.4 trillion from the IMF a few weeks earlier.

Since then we have heard the lame duck President George Bush, at the time of writing, feeling it necessary to take airtime to tell the American people that capitalism is really a jolly good thing – 'I'm a free market kind of guy' – and to stick with it, while the word 'socialism' entered the presidential election campaign. This came after watching the nationalisation of American and British banks by the Bush Administration, and by a New Labour government (one that advertised itself as 'prudent', while selling London as the least regulated global financial centre in the world), who did their utmost to call it something other than 'nationalisation'. In other parts of the world banks are also 'saved' or part-nationalised.

The question then for people who see capitalism as a mean, unjust and crippling political-economic form, is how to use this political opportunity, rather than arguing over any exclusive formulation of a Marxist theory of crisis. One serious concern is media presentation, the framing and presentation of the recession through the fetishised drama of rising and falling stock market figures. The BBC speaks of 'demystifying' what has happened, 'with the experts and correspondents to give you a comprehensive explanation of events in the financial

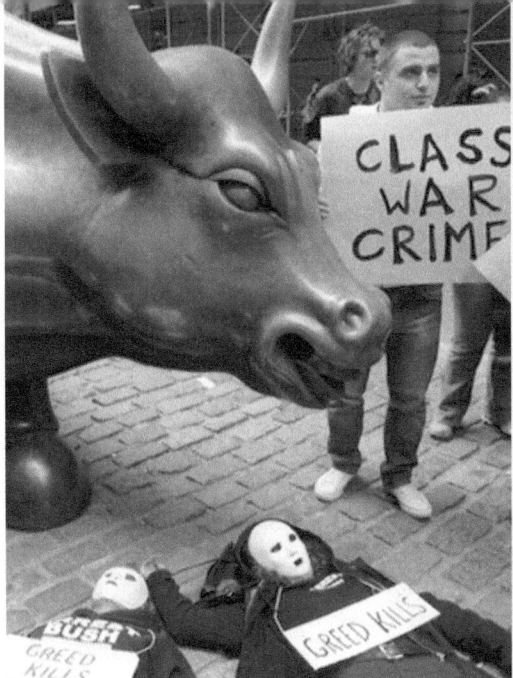
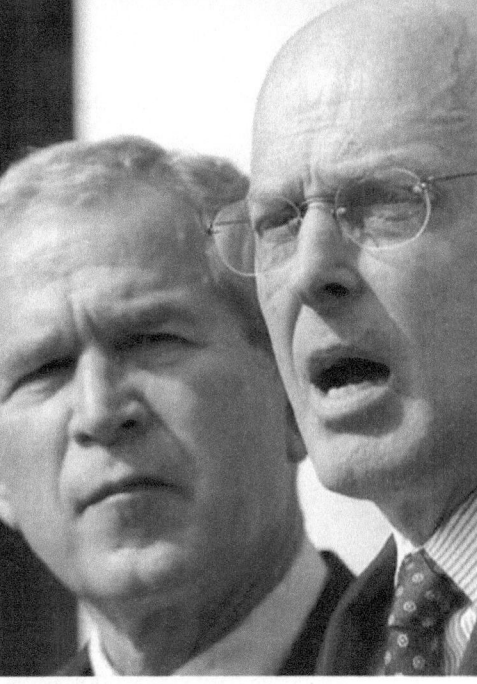

Image: left – protestors at the Wall Street Bull demonstrating against the US government's proposed $700 billion bail of the financial system, New York Stock Exchange, 25 September 2008. Right – former President George Bush and Treasury Secretary Henry Paulson

world'. In fact it is a thoroughly partial explanation; a monologue of many colours which includes the apocalyptic, but is exclusively in the hands of finance world wiseguys, wise-after-the-event wiseguys (bankers, financial journalists and analysts) giving in-house, Green Zone analysis.[1] They speak with authority even when they are incoherent, and when what has happened has shown them to be ignorant or greedy wishful thinkers. Meanwhile the ex-head boys and girls in the studio automatically refer to socialism as 'old-fashioned'.

Over the last year there has been a cycle of denial regarding the 'soundness' of venerable financial institutions and the 'seriousness' of the situation. Ever since US investment bank Lehman Brothers went down the tubes and things got seriously out of hand, the authoritative voice has switched, without shame, to a narrative of catastrophe, but catastrophe as defined in its own terms.[2] 'Staring at the abyss', and 'on the brink', and then 'averting a catastrophe'. These phrases give a fetishised word-picture of 'economic collapse' as a singular, melodramatic event, and have been repeated so many times that it becomes boring enough for most folk to fall asleep while trying to work out what the hell *is* going on. Meanwhile the abyss turns out not to be an abyss, but rather a dragged-out austerity phase of privatised, watching-the-pennies, anxiety for the majority.

At the same time – to scare the children/taxpayers – there's 'meltdown', as in nuclear disaster. It is scaring the children with a purpose: to save the banks, more or less on their own terms, or who knows what may happen in the *real* world. This current Green Zone concern with the *real* economy, ought to imply a recognition that what we've been fed before is a fantasy world, but its functions are very different. Treasury Secretary Paulson, for example, has talked repeatedly of 'using all the tools in the locker' when dishing out more money, as if fixing an engine with a temporary fault. At the same time this *real* economy becomes somehow virtuous, as if it wasn't a business of armaments production, 'terminator' seeds, and the super-exploitation of export processing zones.

Various sensible demands have been made on behalf of the 'taxpayer' as quid pro quo for any bail-out of banks; sensible demands which include, but go beyond, a bail-out tax, or the prevention of foreclosures for mugged house buyers. Sensible demands have also addressed the deflationary recession that is in process. These demands include directing money towards turning empty, or mothballed, new build housing into social housing, raising the minimum wage and increasing social security payments, all of which would increase 'effective demand' to counter the recession.

But such sensible proposals – whether we call them 'reformist' or not – will not have the smallest chance of success if the authoritative voice is not challenged at every opportunity. Understanding that the Emperor is naked is no guarantee of the radical change that present circumstances demand, but is a prerequisite. As it is now, public challenge has been monopolised by useless professional hands, not just useless but part of the monologue. There is the irrelevant irreverence of satire and 'alternative' comedy, the routine grilling of easy-prey politicians, and demands from professional opinionists that capitalism be nicer. When it comes to the economy, what we get are unchallenged experts, all of whom have been party to the free market fantasy world. It may have a coat of many colours, but the monologue is highly selective. So much so that we don't even get to hear trade union leaders on mass media, not even when their quid pro demands for rescuing greedy bankers are as weak as they are. What this suggests, however, is that this neoliberal/free market narrative of brazen capitalism has a thin skin. Alternatives of all sorts, however mild, are seen as a threat.

Crises are moments when bright lights are shone on what was hidden and opaque, all those murky corners that are integral to capitalism. There have been

> **the *real* economy should imply that what we've been fed before is a fantasy world**

many instances of naff opportunism by the British left, but these should not put off people who see and feel capitalism as unjust and archaic from taking this exceptional opportunity to keep those bright lights aimed at its greeds, pretensions, and wishful thinking. It is also necessary to fight off blame being attached to 'illegal' immigrants, generic Muslims and the casting of 'finance capital' as Jewish capital. These possibilities give the job urgency, as does the need to counter Green Zone in-house attempts to explain away recent revelations, keep the free market's pretensions afloat, and use the crisis as a means to accelerate the capitalist process of monopolisation.

Whose Free Market?

Looked at with a cold eye, its pretensions are ludicrous. The reality of the free market has been of 'uneven development' with its historically-loaded dice; of various extremes of exploitation; of a process of monopolisation; a dependence in its 'Anglo-Saxon' heartland on a military Keynesianism; and on the privatised debts and anxieties of large numbers of people. The perverse pride in its impersonal, yet anthropomorphic power – the invisible hand – is laughable given its penchant for gurus. Gurus are always liable to fall from grace, as in the case of Alan Greenspan, but another one is there ready to take his place. This time, Warren Buffet, the *sage* of Omaha, whose investment decisions and even opinions can alter what is grotesquely called 'market sentiment'.

Free market ideology relies on crude notions of inevitability – There Is No Alternative – with its convenient denial of human agency. After the East Asian currency crisis of 1997-8, which free marketeers had blamed on the 'crony capitalism' apparently characteristic of that area, the then guru Greenspan called its resolution – the enforced liberalisation of capital markets in Asia – 'a milestone on the *inexorable* trend towards market capitalism.' Even more crude was a celebration of the ruthless nature of this inevitability. Commenting on another crisis of that time, an attack on the Brazilian currency, David Smith, Economics Editor of Murdoch's *Sunday Times*, wrote:

> It may not always be pretty, but it is the way international capitalism works. Control it or insist on tougher lending criteria and the supply of capital that is the lifeblood of economic development will dry up. You cannot pick and choose, or try to put the genie back in the bottle.

What has now made these pretensions risible is that all the self-styled tough, independent characters who make up this market, having derided governments, and

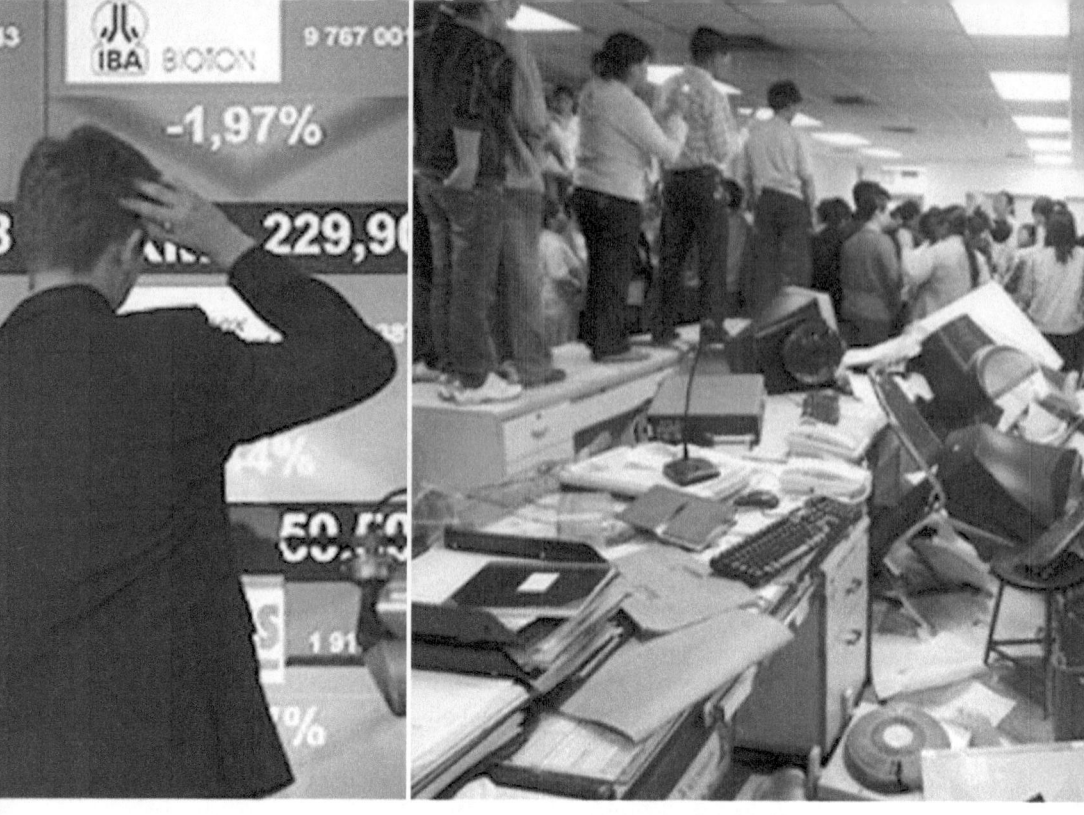

government interference, have had to be rescued by taxpayer-financed governments. It would be inhuman not to enjoy the hubris of the market and its representatives, and the real dent the crisis has put into its fantasy version of itself. Governments and government action were ridiculed because if they started to influence risk and investment decisions they would mess it up; because they were helpless to influence the market due to its sheer size and global nature; and, most of all, because it would give the lie to the fetish of 'the independence of money'.

In an interview two years earlier, city wiseguy Graham Bishop referred to the market as 'a rolling referendum on the views of savers about government policies.'[3] In the same interview he talked of how 'pressures from the owners of securities will stimulate a bottom-up restructuring of European industry' and how this boded well for EU competitiveness. It is pretty clear then what kind of government policies were judged in this rolling referendum conducted by 'market confidence': the relative and competitive conditions made for maximum and secure extraction of surplus value.

'All that was Solid Melts into Air'

The methods by which governments created a situation where 'market' judgements could have such a fast and dramatic impact on their own powers are well known to be the deregulation and liberalisation of finance capital. This was a

Image: right – factory workers occupy an office after smashing equipment during a protest at the Kaida toy factory, Dongguan, Guangdong province, China, 25 November 2008

John Barker

politico-economic strategy beginning in the early 1970s and aimed at those who created surplus value both for individual capitalists, and for capital as a whole with its incessant need to accu-

brazen capitalism has a thin skin. Alternatives, however mild, are seen as a threat

mulate. For its demands to be met the demands of workers, both economic and cultural, had to be disciplined. Deregulated capital fetishised as 'the market' and free to go where it pleased took on this disciplinary role with the great advantage that it could not be negotiated with. After a two month strike at General Motors, a UAW official talked of how it had been impossible to negotiate in this 'universe of shareholders and analysts', one which had replaced that of industrial managers whose authority had been undermined by labour militancy and the cultural confidence that went with it. This reassertion of the power of capital did not take place in isolation. The skewing of infrastructural investment towards telecommunications and information technology coalesced with the dynamics of transnational corporations.

This macro-role in the push to extract more surplus value was matched by a micro process of mergers and acquisitions, and lately by the managerial activity of private equity, all credit financed. It meant a hands-on role for finance capital in creating more surplus value in productive sectors of whose material processes the new 'managers' knew nothing. All they know is to increase the intensity of labour, that is, the same or more work done by less workers. But both at micro and macro levels, and taking into account the huge increase in the global extraction of surplus value coming from South and East Asia, the global pot of surplus value is always finite at any given time. Needless to say, however, the market, active in a range of price movements of every conceivable currency or commodity, material or 'immaterial', developed its own set of interests in increasing its share of that finite pot to beyond what was possible.

The self-regulating, 'invisible hand' market proved to have partial vision, prejudices and irrationalities. Its collective blindness was being unable to see that the pressures it exerted in surplus-value extraction, whether it be union-breaking, relocation or job cuts, meant lower real wages across the board, and this would create problems in the realisation of surplus value. Ten years ago wiseguy Ed Yardemi was already saying that the world needed all the yuppies it could muster. There have not been enough and, though a relatively large-scale bourgeoisie is being created in China and India, their consumer spending power is still relatively small. Instead, as has been obvious for several years, the circle was squared with a rapid increase in personal debt, which in the UK began during the time of Margaret Thatcher.

That especially partial vision of the free market chose to ignore the fact that this could not last forever. Faced with insufficient world-wide numbers of yuppies, but with an addiction to high-rate-of-return accumulation, capital turned to ripping off the poor in the sphere of its own reproduction. It survived in the meantime by both creating and being part of a fantasy world, one in which house prices, for instance, would never fall. Back in 1998, 'market sentiment' had created the reality of a Brazilian currency crisis, despite what used to be known as economic 'fundamentals', like its Balance of Trade being strongly in surplus. This seems to have lead to the belief that all 'fundamentals' had been somehow overcome, as with Greenspan's 'new era economics'. It is a DIY reality world which was epitomised by Mr Tony Blair saying 'I only know what I believe' when confronted with unambiguous evidence of the absence of WMD in Iraq.

Sound Bite History Repeats Itself

A basic requirement of a fantasy world living in an eternal present is not just to ignore, or not see, awkward facts, but to blank out history however recent. Around ten years ago there were crises which, though small scale looked at from now, were not seen as such at the time. The East Asian currency crisis, followed by a Russian debt default and the subsequent collapse and rescue of a major hedge fund of the time (the ironically named Long Term Capital Management), caused a brief period of self-examination in the financial world. Then there was also talk of 'tottering on the brink' and 'infection spreading'. President Chirac proposed reforming the IMF and annoyed the Americans as Sarkozy is doing now. There was talk of a global financial architecture by many of the same people talking of it now, like George Soros, and also the *Financial Times*:

> LTCM shows that it is not only in developing markets that transparency, oversight and prudential controls have been found wanting, A growing proportion of global capital is leveraged, so the G7 needs to make progress on two fronts next week. It must start to create a global financial architecture that can deal with problems in emerging markets. But it must also deal with problems of transparency controls, risk management and regulation much closer to home. (*Financial Times*, 26 September 1998)

That was ten years ago, and as we know, none of these things happened. The whole idea was rejected by the Bundesbank even more vehemently than the US Treasury.

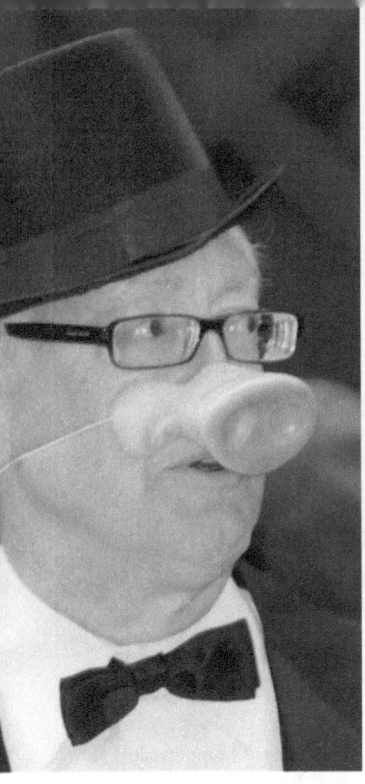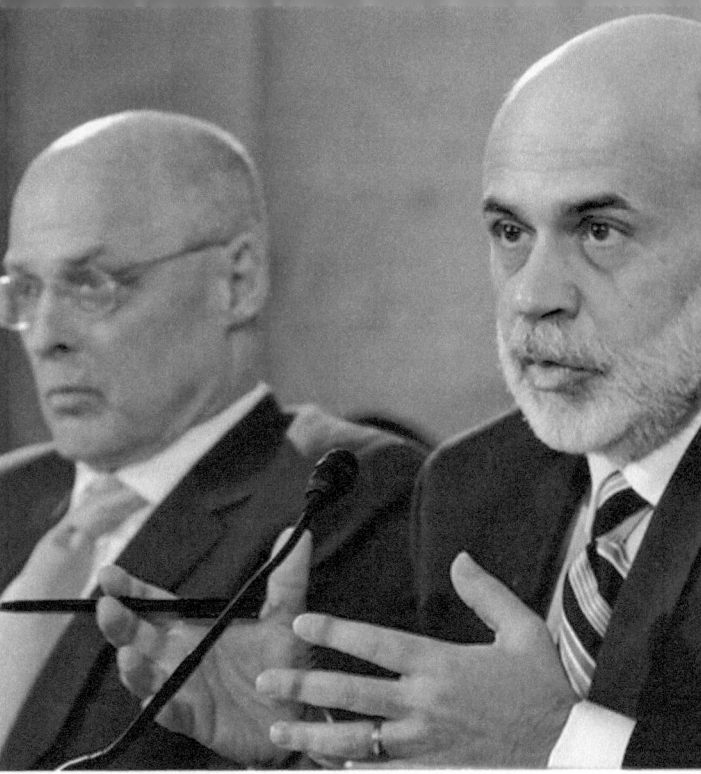

Other headlines of the time sound equally familiar:

> The situation raises questions of banks' risk management practices, but the lure of high fees on derivative products and the increasingly lax lending standards of recent years contributed to the willingness to take risks. (*Wall Street Journal*, 2 September 1998)

> Greed has replaced bankerly caution, this under the misguided assumption that some players were too big to fall. (*Herald Tribune*, 7 September 1998)

It is even more eerie, in light of Paulson's $700 billion bail-out that was pushed through Congress without proper debate, to read what was said earlier in that year of 1998:

> Before pumping more money into the system, Congress has the right to ask what's being done so that it will not have to choose again between aiding undeserving bankers and risking a global collapse. (*Washington Post*, 29 January 1998)

The language at that time was also similar, the collapse of the *real* economy was the alternative, an apocalypse. The difference is that in 2008, disaster rhetoric came into use after tens of billions had *already* been provided to the banks by the Fed and the Bank of England against collateral of uncertain value, and this with the promise that said billions would *prevent* apocalyptic circumstances. What the similarities do suggest is that capitalism could not reform itself, even in its own

Image: left – anti-US bailout protestor. Right – Fed chief Ben Bernanke testifies during a Senate Banking, Housing and Urban Affairs Committee hearing, 14 February 2008 long term interest. The dotcom bubble was followed by Enron, while the only significant US legislation of the time was the Sarbanes-Oxley Act, Rule 46-R which allowed for those notorious off-balance sheet special entities.

This time the crisis is on a massively different scale, but that has partly been caused by what was supposed to be the safeguard to Rule 46-R, that these off-balance sheet vehicles were OK so long as the bulk of reward and risks lay with others. That assumed financial institutions would be wholly truthful to each other.

'When I choose a word, it means just what I choose it to mean' (Humpty Dumpty)

This assumption has been blown away by the inescapable fact that banks are reluctant to lend to each other, knowing themselves what can be done with balance sheets and asset values and how much deception they are capable of; blown away, too, by the sight of financial institutions suing each other for having been misled over the value of the various financial packages passed on (i.e. how secure the income streams on which they were premised actually are). Big name institutions, that have come on like wronged virgins in this pass-the-parcel game, are then sued themselves for having shafted smaller funds.[4]

Within the Green Zone, this has exacerbated a concern as to the 'small' investor's loss of faith in the market which had surfaced during the dotcom bubble, so much so that in June 2008 two of the big honchos of the collapsed Bear Sterns were indicted not just for routine 'insider trading' but 'for not giving *full* information to investors'.

Outside of the Green Zone, what the crisis spotlight has revealed is that in the fantasy culture there is only partial information: one where objective scientific research turns out to be funded by corporations with a distinctly partial interest in the results. In short, it has shown what was already there, a crisis in the integrity of information. This is especially serious in what it says about the parallel fantasy world of the 'information society'. Marcel Rohner, the new CEO of UBS Bank – which, though both virgin and seducer in this melodrama of mis-selling, still made heavy losses – rationalised what had happened by blaming not the lack of integrity, but rather the overload of information. This is hardly new – the CIA started the trend when explaining its failure to know that the first Indian nuclear bomb test was about to take place – but Rohner's is especially revealing:

> The problem was not a failure to appreciate complexity, but rather the opposite: it was a lack of simplicity and critical perspective, which prevented the right questions from being asked while there was still time.

John Barker

This is another version of 'bankers ain't what they used to be' which has been heavily used in capital's obfusctaing self-explanation, as in, 'it's all the fault of

did anyone in the fantasy world actually believe the poor would only get richer if the rich got richer?

those mathematicians in the financial world.'[5] But it does undermine the rationale of the Information Society, which has not had 'critical' thinking high on its agenda for non-elites. Even in universities, critical thinking has often been silenced by the exceptionally partial analysis of unchecked think-tanks, fanatics of the fantasy world, and the publicity they are given.

The complexity Rohner describes, and perhaps the very notion of such an information society, involves ever greater levels of abstraction and this, along with the comprehensive predominance of marketing and its language, is a characteristic of the fantasy world. It is a continuation of the dynamic by which exchange values came to dominate use values. It was highlighted in the recent crisis by the estimate that Deutsche Bank had become the biggest landlord in Cleveland, Ohio, with plenty of empty properties to rent. It is hard to believe that any executives or the risk committee of the bank had ever been anywhere near the streets of Cleveland. Living in another world in which there are no queues or weekly anxieties, these bankers appear not to have been informed that real wages in the USA had been stagnant or falling in the last 10 to 20 years. Things were going well for them, it followed then, as if they actually believed the 'trickle-down' narrative, that *everyone* was happy.

Such abstraction is fertile ground for the capitalist fantasy world. A *Herald Tribune* headline (12 June 2008) read 'Housing is booming if only on television'. The audiences for HGTV and TLC, the two US networks with the most 'property programming', it reported, have been growing over the last three years. As the housing market slumped, the scheduling of *House Hunters* and *Designed to Sell* increased dramatically. R. J. Cutler, a wiseguy producer of one of them, *Flip the House*, commented with a routine piece of bogus analogy, that 'People had loved comedies during the depression too'. It is as if, along with the credit, more fantasy had to be 'pumped' into the system.[6]

Eyes Wide Shut

The crisis of information integrity for the capitalist class takes the form of a breakdown in the objective evaluation of assets. Assets with a face value of mil-

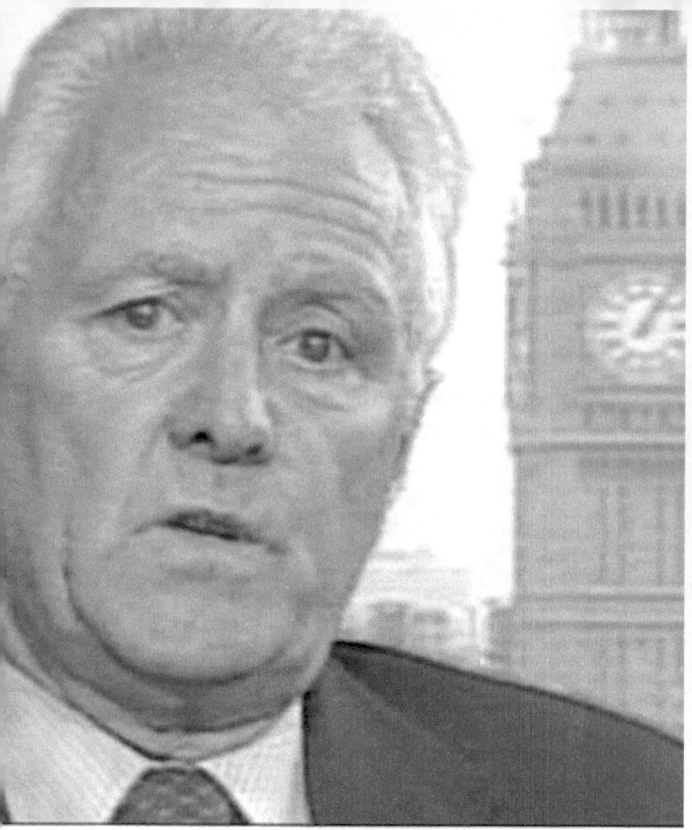
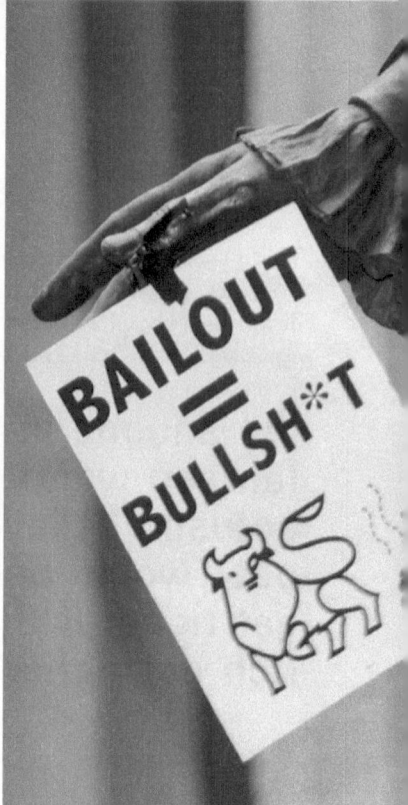

lions turn out to have little or no value because they were dependent on a stream of income, the source of which capital itself had already squeezed. All this seems surprising given the cultural obsession with numbers in American society, and when the mantra of the ubiquitous McKinsey Consultancy, pioneers of securitisation, is: 'if you can measure it, you can manage it.' There are private profit-making 'institutions' that were supposed to assess values – ratings agencies and auditors. Given their form, it is amazing that the oligopoly of global auditors has continued untouched in this role. Take Coopers & Lybrand with its auditing record for Robert Maxwell, Polly Peck and Barings Bank. In light of these successive scandals, it took the very British step of changing its name by merging with Price Waterhouse, and, in the process, furthered the process of monopolisation. It now turns out that after the notorious case of Arthur Andersen's complicity in the frauds of Enron and the promises to police conflicts of interest in the future, PriceWaterhouseCoopers were auditors *and* consultants for Northern Rock, whose collapse and the shocking sight of queues of people at its doors demanding their money, visually symbolised an end to the fantasy world.

Ten years ago – once again – the analysis was there in respect of ratings agencies: 'What were the banks research departments saying six months ago? Nor did the IMF or ratings agencies such as Moody's and Standard and Poor, provide any warnings...' complained the *Washington Post* (6 January 1998). Fitch's makes up the cosy threesome oligopoly of these agencies, and they too have happily managed not

John Barker

The ratings agencies' role is crucial to sustaining the fiction of the independence of money

to bother about their conflict of interests, not when it comes to the creditworthiness of banks, their assets and financial 'products', which is what they are supposed to be objectively rating with their triple As and so on. Suddenly it is common knowledge, as if we'd known all along, that these agencies were being paid by the financial institutions whose bonds and assets they were rating. And – this is how brazen one can be in a fantasy world – the banks could pick and choose among the three to get the best rating for their money.

It is crucial that the bright light remains on these sectors that the crisis has brought out of the shadows. Once the Green Zone's smothering strategies (and its determination to control interpretations of what is happening) was succinctly described in the *Herald Tribune* (5 June 2008): 'Regulators have struggled to assign blame for the mortgage debacle, at times pointing to everyone and no one.'[7] While it has been entertaining to watch the 'masters of the universe' blaming everyone else including each other – it's not me guv – it is important not to let this diffusion of blame become like the market itself, where responsibility cannot be pinned on anyone in particular or any of capital's own service industries.

Secondly, both play crucial roles in capital's version of itself. The oligopoly of auditor consultants are prime movers in the murky business of carbon emissions trading, and in the global and profitable business of public-private partnerships and direct privatisation. The ratings agencies' role is especially significant as a means of discipline in which the fiction of the independence of money can be maintained and, as a corollary, the unique efficiency of private capital in making investment decisions. Speaking at the time of that Indian nuclear bomb test which the CIA had failed to predict, Felix Rohatyn spoke of the disciplinary power of the market, how it had punished India in a way diplomacy could not:

> While many countries refused to sanction India as a result of its nuclear tests, the capital markets provided that sanction promptly.

But this required the mediation of Standard and Poor, who, he went on to say,

> downgraded the outlook for India from 'Stable to Negative', thereby raising India's borrowing costs immediately. The Bombay Stock Exchange slid and the rupee lost 10% against the dollar.[8]

This was not the market itself, but a political decision, masquerading as non-political. Sometimes it has worked the other way. At the time of the East Asian currency crisis the agencies, who had seen nothing wrong with its economies, followed the market when its 'sentiment' turned sour, and by doing so amplified the misery that ensued in that part of the world by raising the cost of credit.

Reality Check

It's hard to believe that anyone in the Green Zone actually fell for the trickle-down narrative in which the poor would only get richer if the rich got richer, though all subscribed to it – smug journalists, think-tank fanatics and its multi-disciplinary marketing department. The real 'fundamental' development in the fantasy world period is one of a steady increase in inequality, and a spectacular increase in the proportion of global wealth taken by the very rich, the top 1 percent. In the USA, this 1 percent took 9 percent of national income in 1979, and by 2005 it was 17.5 percent. Their greed is at the peak of a coalescence of personal and family greed which has added another dimension to capital's compulsion to accumulate. By the simultaneous pushing down of real wages and the claims on the global surplus value exceeding its generation that this involves, an economic crisis has ensued. Our job is to make of this a political and cultural crisis.

A couple of years after 1998, all the talk of financial architectures, the greed of banks, and reform was seen to have had no consequences. This too is typical of the dark, marketing side of the fantasy world in which promises of disaster aid are made but not actually delivered. By then things had also quietened down in the fantasy world at the expense of the poor people of Indonesia and other South East Asian countries whose suffering was not spectacular enough to excite much comment. It meant that the *Financial Times* could adopt a blasé voice regarding convictions made a case of share price manipulation:

> Business morality, like business itself, is cyclical in nature: during periods of financial euphoria and strongly rising share prices, people cut corners and bend the rules; during the austere times which follow, the rule books are rewritten and everyone agrees things will be different next time. (*Financial Times*, September 1990)

John Barker

It's as if the cyclical nature of the economy was a comfortable fact – unsurprising when austerity has no impact on its writers. Our job is to make sure that the sobering up process is in our own hands and that it exposes the fantasy world and what it has hidden.

This time there is such an opportunity because things are rather different (and, it should be emphasised, still in process with another huge bail-out of Citigroup) not just because of the scale of what has happened, but because the free market has had to run to the state and thus to 'the taxpayer'. It is an opportunity to escape the separating roles that are imposed on us as taxpayer, consumer, and – the ghost in the cupboard – producer, to become class-conscious citizens, a not-impossible process when so many types of work have been proletarianised.

What will not help is to get stuck in the coils of what is 'reformist' and what is 'revolutionary'. What matters most is *how* gains are fought for, and the dynamic they engender, which have the possibility of going further than their representation. Otherwise we are likely to be paralysed by these ahistorical concepts, which have been further confused by the monopolisation of the very word 'reform' by the useful idiots of the free market. 'Reformism' is often understood as the working class being 'bought off', as if the class were saints in overalls being led into temptation. That is both moralistic, undialectical, and – in the context of modern globalisation as opposed to previous periods of imperialism in which cheap imports were not balanced against falling real wages in the western world – ahistorical. The evidence from ten years ago is that capitalism has great difficulty in reforming itself because there are basic psychic fears in the mean and foxy capitalist mindset of give-them-an-inch-and-they'll-take-a-mile, of being drawn into anything that compromises the inalienable right of private property, and even of the transparency it espouses. This can be seen in the secretive distribution of the October Paulson hand-out by an ad hoc body. Keynesian economics is 'reformist' in that it is premised on exploitation, but it is subversive in its emphasis on human agency as against an omnipotent market. The stark overturning of this ex-reality is there for all to see, and shows that the possibility for radical change does exist.

On Alert

'Reformism' should not be confused with the achievement of *all* limited class gains. Some reforms defuse popular anger, others give it focus, as for example those that concern housing and its class-skewed structure which at present is wholly to the benefit of a rentier class, banks and privileged individuals. It is the defusing of anger which characterises reformism, when the maximum gains possible against a weakened enemy are not made, or even demanded, or smothered by irrelevant modifications. This is a danger, especially when the capitalist class wants a period of consolidation of its gains, a quiet period when it will be content with much lower rates of return: the wholly self-absorbed 'flight to safety'. Alertness is required, rather than the paralysis brought on by a fear of 'reformism'.

The presence of popular anger as a political force – so easily dissipated, as George Caffentzis describes after Paulson's bank-friendly bail-out[9] – instead requires, amongst a variety of tactics and means of opposition, an immediate slapping down of:

> -all attempts to create new fantasies, like the idea that 'the taxpayer' may profit from the bail-outs;
> -the notion that only capitalism can allocate resources efficiently, or rather that it does so at all, and in the process raising the question of 'efficiency for whom?';
> -the idea that finance capital is a temporary excrescence as against the 'real' economy presented as virtuous and wronged;
> -the fantasy that bank investment can be directed without nationalisation. (In this instance such a slapping down does not mean necessarily that nationalisation should be demanded, just that a new fantasy is being proposed);
> -that capitalism can ever be stable, let alone just.

It also means not being afraid of popular anger. Capitalism is not a 'system' but a mode of production which requires its own self-interested agents. It's one of Marx's great strengths that he is both moralist and analyst, and that while unpicking capitalism's dynamics, he names names, attacks particular individuals and particular institutions which today include accountants and ratings agencies. Nothing seems to have angered people more than the fund manager who said that banks could not hold down pay and bonuses, because it would cause 'an exodus of staff to Mumbai, Shanghai or Dubai'. The response was, 'Go on then', an instinctive attack on the apparent expertise that is a fundamental justification for massive inequality; and on the brazen shamelessness of bankers being expressed by the ways in which taxpayers' money is being used to concentrate finance capital by merger and acquisition, for example; and in the wholly non-transparent manner in which Paulson's $700 billion is being distributed.

Our job is to make of this a political and cultural crisis

John Barker

Against this shamelessness, a whole language of contempt is required. One that has to storm the Green Zone of capital's own explanation of what has happened and what will be done; one that does not permit the smothering of the rebuke 'Private Profits: Social Losses'.

Footnotes

1 The Green Zone is the common name for the International Zone in Iraq [editor's note].

2 Why then was Lehman Brothers allowed to fail when Europeans like the French Finance Minister, Christine Lagarde, saw this as the trigger for the global crisis? At the time, Ben Bernanake said: 'A public sector solution for Lehman proved infeasible, as the firm could not post sufficient collateral to provide reasonable assurance that a loan from the Federal Reserve could be repaid. And the treasury did not have the authority to absorb billions of dollars of expected losses to facilitate Lehman's acquisition by another firm.' This, as various people have pointed out, simply doesn't add up. Not when the Fed had proved willing to buy so much debt without collateral, and when guarantees on liabilities were given in the case of Bear Sterns and Wachovia. One can imagine a conspiracy theory in which it was all an American plot to pass the parcel. More likely, given their record elsewhere, the authorities simply didn't understand the consequences of allowing Lehman to fail. Since then blame has been narrowed down to the bank's CEO, Fuld, for trying to drive too high a bargain in finding commercial partners or buyers. None of which prevented him personally making some $480 million over the last ten years, and then, when questioned about it in the House of Representatives, coming on like a Holocaust denier and saying that really it was only $400 million.

3 Robert G. Williams, *The Money Changers*, Zed Books, 2006.

4 UBS are in the middle of this. The original impression given was that they were one of the mugs in a game they didn't understand, but in February 2008 they were sued by, and then counter-sued, the Paramax hedge fund for apparent misrepresentation by the bank. It also has to fight against the German HSH Nordbank for 'mis-selling and misrepresentation of risk.' The volume of such cases has provoked a book called, *The Pebble and the Pool: The Global Expansion of Subprime Litigation*, John Doherty & Richard Hans, Tomson West 2008.

5 See John Barker, 'Structural Greed', *Variant*, issue 32, Summer 2008.

6 And as if Preston Sturges' great movie, *Sullivan's Way*, could ever justify this kind of banal fantasy.

7 This singular blame on subprime has itself been deceptive, and part of the smothering tactics used in media presentation. More exactly it was the straw that broke the camel's back loaded down with leveraged assets. See Barker, op. cit.

8 Rohatyn, New York's financial 'saviour', then US Ambassador to France, but lately of collapsed Lehman Brothers. Barker, op. cit.

9 'Notes on the Bail-Out of the Financial Crisis': http://news.infoshop.org/article.php?story=20081022233913679

John Barker <harrier@easynet.co.uk> was born in London and works as a book indexer. His prison memoir *Bending the Bars: Prison Stories* was published by Christie Books

THE WHO AND WHOM OF LIBERTY TAKING

The British Library's show Taking Liberties – on the struggle for freedom and rights throughout 900 years of British history – is a welcome event, writes Peter Linebaugh. But, he wonders, is it possible to discuss liberty while excluding the crucial question of equality?

Trying to make my way to the 'shrine' in which Magna Carta is presented, I hover at the edge of a group of school children taking some liberties themselves – giggling and chatting while huddled on the floor to listen as their teacher prepares to offer them lessons in earnest. Mounted above is a quote from John Locke, the bourgeois theorist of competitive individualism and private property, 'Where there is no law there is no freedom'. This is the note, at once abrupt and severe, upon which the exhibit begins. Clearly, this is an exhibit with an argument. (Hush now, children.)

Let's look at Locke more carefully.

> For in all the states of created beings, capable of laws, where there is no law there is no freedom [...] freedom is not, as we are told, liberty for every man to do what he lists (for who could be free when every other man's humour might domineer over him?), but a liberty to dispose, and order as he lists, his person, actions, possessions, and his whole property, within the allowance of those laws under which he is.

This was written in 1689 against the English revolutionary antinomians, diggers, levellers, those without money and those who'd turn the world upside down. They were familiar with custom and solidarity. They did not do with their persons, property, and possessions as they list, but laboured to subsist with others. The humour of woman and man may be to cooperate. For truly the stance to law was different for the commoner (s/he ducks) than the bourgeois citizen (s/he puffs), as explained in Christopher Hill's powerful book *Liberty Against the Law* (1996). As the attitude to law differs, so does the take on liberty. Thus warned, I advance into the exhibit.

Peter Linebaugh

Wow! There is Magna Carta! The medieval Latin, the minuscule clerical writing, the dim lighting are all too much for most of us, obscuring what the exhibit as a whole wants to be made legible, visible and comprehensible. A banner above Magna Carta quotes chapter 39 in English translation ('No free man shall be seized or imprisoned [...] except by the lawful judgement of his equals or by the law of the land'). A translation would be a welcome gift of the library to the patrons, in keeping with the wired wristband you pick up at the beginning to permit you to record your own opinions in the interactive consoles as you make your way from one group of exhibit cases to another. Fake, crumbling cement walls partition these different areas with the reinforcing rods showing through as if the walls were recently bombed or destroyed. Coming as I do from the USA, I think of Fallujah or our domestic gulag which is greater than any other country in the world proportionally to the population. Despite chapter 39, prison destruction is not a theme.

This is an important, interesting exhibition, and however you may stand on the arguments implicit and explicit in its organisation and selection, it is worth a visit and costs nothing. There is an amazing variety of objects on display – parchments, seals, jugs, truncheons, a purse, a funeral card, a diary, a notebook, a medal, a doll, paintings, cockade, banners and flags. Many, many items signifying liberty. The indirect ones hit hardest, like Emily Davison's unused return ticket to London (she had not planned to be crushed under the hooves of King George's horse).

It doesn't take long to simply walk through the exhibit. The areas are marked: 1) Liberty and Rule of Law, 2) Parliament and People, 3) Right to Vote, 4) United Kingdom?, 5) Freedom from Want, 6) Human Rights, 7) Freedom of Speech and Belief, 8) Interactive Results and 9) Your Thoughts. The more you pause and study the more you enter into the labours of history.

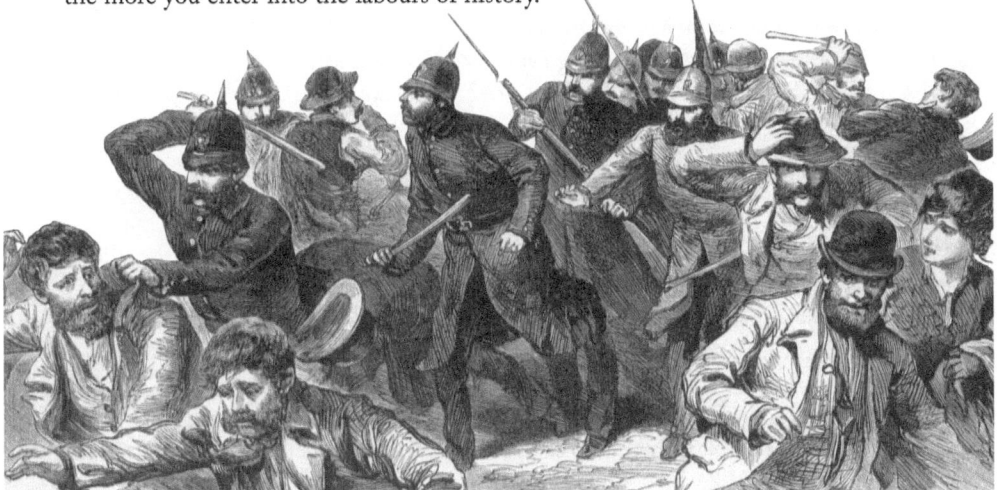

Image: 'Police and Hussars Charging the Proclaimed Meeting at Ennis: Scene in the Courtyard', from *Illustrated London News,* 21 April 1888

The great documents of the 17th century are here: the death warrant of Charles I (1649) signed by 57 regicides; the Petition of Right (1628); Coke's *Institutes* (1642) with their revolutionary theories of law; the notebook recording the Putney Debates (1647) and the egalitarian cry of 'the poorest he' against the fears of property; the Agreement of the People (1649) is a huge document almost from ceiling to floor; the Habeas Corpus Act (1679) with its explicit condemnation of overseas rendition of trials and torture. These are the treasures much revered by the early working class movement of the 19th century.

A tone of smugness mars the exhibit. The jetlagged traveller making his way from Heathrow peers out the window of the tube to see the poster in the passing Underground station, 'In some Countries you wouldn't have the right to visit this exhibition about your rights.' What humbug! Unless the exhibit is going to travel? Is it a reference to China? Very early in the exhibit there is the story of Sun Yet-Sen and his detention in the Chinese legation until a British judge, citing Magna Carta, ordered his release in 1896. These documents have become part of a world archive. After the Americans permitted the destruction of the archives and library of Iraq containing humanity's record of the first cities, we're reminded that the British Library, Linda Colley puts it, 'is an institution that calls itself British but which belongs in fact to the world.' Despite occasional self-satisfaction, Professor Linda Colley, the curator, and her colleagues are to be congratulated for bringing these treasures to light from the dust and dimness of the Bush-Blair years.

the advocates of freedom will always seem to be clown- ish or rude, at least to begin with

The title of this exhibit, Taking Liberties, goes back to 1980 and a famous quotation by E.P. Thompson, the social historian and peace activist. 'For two decades,' he wrote, 'the state, whether Conservative or Labour administrations, has been taking liberties, and these liberties were once ours.' England had the most reactionary government in Europe he argued, 'simultaneously attacking the livelihood and democratic rights of its own people.' He referred to spies, data collection, jury vetting, surveillance state, interceptions of mail, tapping of telephones, trolling internet traffic, and police murder. This was the 'secret state' (as Thompson called it) which looked back to 'Old Nobodaddy' (as Blake called it) or 'the Thing' (as Cobbet did) and forward to the world of Closed Circuit Televisions (CCTV), Anti-Social Behaviour Orders (ASBOs) and High Net-Worth Individuals (HNWIs).

In the phrase 'Taking Liberties' Thompson introduced something powerful, a double entendre mixing state crimes with bad manners. 'The taking of liberty'

Peter Linebaugh

transgresses the bounds of propriety. A.A. Milne's children's poem explains:

> Excuse me, Your Majesty
> For taking of the liberty
> But marmalade is tasty if
> It's very thickly spread.

'To go beyond the bounds of civility', yes, that definition pertains to the subject since the advocates of freedom will always seem to be clownish or rude, at least to begin with, their voices too loud, breaking the polite murmur of civil society, or their elbows and knees knocking over the marmalade, jam, honey, and all the breakfast dishes. (Walk Don't Run, children.) The exhibit is firmly within bounds. Its icon is the clenched fist of working class solidarity – 'one and all, one and all' – rendered here neither in anarchist black nor communist red but in pastel hues of pink or turquoise. The dangerous gesture has given way to an attractive brand.

The phrase catches us, we want to go along with it, but as we think about it, it becomes one of those brain teasers that tugs in opposite directions. As a gerund it does not distinguish subject from object. Is it about attaining liberty or getting rid of it? Lenin summarised the class approach to social analysis by two questions, Who? Whom? Who takes liberty from whom? Thompson and Colley pull the phrase in opposite directions. Thompson says the state takes from the people while Colley says the people take from the state. Hidden in this argument is an academic quarrel going back a few decades.

In a nutshell this is it. In 1963 Thompson published *The Making of the English Working Class* setting a scholarly agenda with 'class' in the middle of it. Colley's *Britons* (1992) put paid to that, placing 'nationality' in its stead as the central problem of modern British history. Thompson had downplayed women and people of colour. Supported by Harvard, Yale, and Princeton in the US, and by Oxford, Cambridge, and the LSE in England, *Britons* was an establishment book. Indeed Colley lectured on 'Britishness' at No.10 to Tony and Cherie Blair as part of the Millenium lectures. Many people took 'class' as a mark of psychological or biological identity rather than as an economic relationship based on material fact, or as an ideal vector of social equality and true livelihood. Thompson increasingly soured by events, no longer found that such a historical class with such an historic task had much meaning, and, as if the idea of class was merely a perishable commodity on the shelf, he announced that it had passed its sell-by date. As an academic quarrel, the laurels fell to Colley. And yet state surveillance mounts, the wars expand, the prisons increase. After all, she returned to the subject and to the phrase.

Her preoccupations remain the same – the franchise and the vote – and with the political entities comprising, what shall we say? Great Britain, or the UK, or England-Scotland-

The Who and Whom of Liberty Taking

Wales-and-Northern-Ireland. Whatever. She is fond of the non-political locution, 'these islands'. There is an exhibit of flags, and the first ideas for the Union Jack in the designs proposed by the Earl of Nottingham under James I. 'Liberty' entails some of the constituent parts of Britain.

For Thompson the problem of identity was always one of politics, neither gender nor 'colour'. As he saw it, there were 'the swaying to-and-fro motions of social class', of privilege and property against liberty and equality. As to identity he expressed a surprising view: 'Take the jury away and I would face a crisis of identity. I would no longer know who the British people are.' Britishness to him was directly related to democracy: 'The jury is, I think, the last place in our institutions where the people – any people – take a hand in "administering" themselves.' The jury, to him was both the palladium of liberty and a random equaliser for the future. The jury system is, he wrote, 'a lingering paradigm of an alternative mode of participatory self-government.' The exhibit alludes once to the jury.

My favourite is the case on the Rights of Man, for here is William Blake's notebook, and his tiny handwriting, with drafts of both *London* (the harlot's curse, the blood down palace walls) and *Tyger! Tyger!*, and the tremendous revolutionary human energy released throughout the Atlantic during the 1790s, and with a pencil sketch portraying Tom Paine. In the same case is a spy's report to the Home Office written after following Paine to Dover. One of Gillray's satirical prints depicts *Patriotic Regeneration* from 1795, as hundreds of delegates with many black faces look on, wearing the bonnet rouge, as the tribune issues out the call to equalise property. Communism and the black faces of successful slave revolt were united in the imaginary of the bourgeois philistine. As absent as the slave revolts is republican Ireland, particularly the United Irish and its revolt of 1798 that directly led to the opposite of a republican independent Ireland, namely the 'United Kingdom'. Over this case, like a foreboding angel, Opie's beautiful portrait of Mary Wollstonecraft has her in her house cap, with the soft hues and light offsetting those penetrating, deep pooled eyes of acute observation and integrity of vision. Framed in ornate gold, the portrait contrasts with her gravestone only a ten minute walk from the British Library in old St. Pancras church yard where it is covered with moss, velvet to the feel of the passing hand.

The affirmative traditions of radical, socialist, communist and labour politics have vanished. Socialism is not here, communism is an absent ghost, anarchism unmentionable, while heresy is a faint wisp. Big themes and ideas such as Class, Equality and the Commons are missing. Then there are incidents and persons absent such as the Peasant's Revolt of 1381, the English Bibles from Wycliffe to Tyndale, the Geneva Bible. We find John Lilburne but not Gerrard Winstanley whose ideas Locke opposed. Where is Jack Cade? Where is Robert Kett? Tom

Peter Linebaugh

Paine but not Tom Spence. Charles Dickens but not Charles Marks. When the census taker came knocking at the door of number 28 Dean Street, parish of St. Ann's, Borough of Westminster, and found the German revolutionary exile, Karl Marx, he recorded him as Charles Marks and so, by the official act of surveillance and data collection, inscribed the revolutionary's name in British spelling. Why is he not here, Karl or Charles? He should be.

Karl Marx wrote of the Ten Hours Act (1848),

> In place of the pompous catalogue of the 'inalienable rights of man' comes the modest Magna Carta of a legally limited working day, which shall make clear when the time which the worker sells is ended and when his own begins.

The workers put their heads together and as a class compelled the passing of a law which prevented them from selling themselves and their families into slavery and death. Marx provides numerous footnotes to factory inspectors; it is not only jurists who interpret this Magna Carta. The chapter resolves the central paradox of political economy which can be expressed as follows: How can labour be both a commodity sold at its value and the source of a surplus value greater than it itself is worth? The answer entails the history and the logic of the transition from the commons to the wage. Serfdom arose from the *corvée* not the other way around.[1] Part of the land was cultivated in severalty (i.e. private property) and part in common, the *ager publicus*.

The affirmative traditions of radical, socialist, communist and labour politics have vanished from the exhibition

Clerical and military dignitaries usurped this land and the labour spent on it. 'The labour of the free peasants on the common land was transformed into *corvée* for the thieves of the common land', writes Marx, and from the *corvée* to the struggle over the length of the working day.

Neither the commons nor equality are themes in this exhibit. Let's think about it for a moment. Can you have liberty without equality? The answer is certainly yes if liberty means liberty of property, then we have rampant privatisation, free trade, inviolable contract. But suppose liberty meant 'the freedom of just conditions' for which the rebel leader Robert Kett is nobly remembered at Norwich Castle? Or, suppose it meant an end of servility, deference, doffing of caps, tugging of forelocks, curtseying? Then liberty entails access to equal subsistence.

As for the commons, in a corner nook in the darkest part of the exhibit, right after you've passed the plastic box with Magna Carta in it, there is the precious, little known Forest Charter. The display case caption reads:

> England's forests had once given the common people somewhere to forage for food and fire wood and space for their animals to feed: the Charter of the Forest restored the traditional rights of the people, where the land had once been held in common.

This can light the lamp of history and cast its luminosity into the darkening corners? We have some of the rights in the exhibit and none of the livelihood. There are liberties to be had as the exhibit shows; there is livelihood too which the exhibit does not.

Commoning has been there all along. 'Commons is become a king', Kett's rebels said in 1549. The fellowship of mutual aid characterising the village community that R.H. Tawney called 'a little commonwealth' or 'practical communism'. Woody Guthrie, the Dust Bowl troubadour, 'When there shall be no want among you, because you'll own everything in common. That's what the Bible says. Common means all of us. This is old Commonism.' In 1941 the Common Wealth Party began to win by-elections with the slogan Common Ownership. It reached its apogee in the Beveridge Report of 1942 with its promise of 'cradle to the grave' social security. John Wycliffe translated Acts 2:44 as to 'hadden alle thingis comyn'. The early Christians 'had all things common [...] as every man his need', a view that entered Marx's definition of communism as, 'From each according to their ability, to each according to need.' 'Where is Fair Shares?' the once defining Clause Four of the Labour Party's constitution stated,

> To secure for the workers by hand or by brain the full fruits of their industry and the most equitable distribution thereof that may be possible upon the basis of the common ownership of the means of production, distribution and exchange.

It is gone from the Labour Party but why does it have to go from history? No, said Thompson, 'there is not a tested prototype of a democratic commonwealth anywhere in the world', and yet this has been precisely the conversation among the world's Have Nots since the early 1990s.

The coal miners strike of 1972 reminded Thompson of 'the egalitarianism of necessity', just as their strike of 1893 reminded William Morris of 'a condition of practical equality of economical condition amongst the whole population'. The energy workers bring this huge awareness of equality and 'the intricate reciprocity of human needs and services'. The relationship between liberty and equality, the relationship between class and freedom, stands at the axis of history. The levellers got their name not as a proper noun, a brand name, but as a common noun with lower case letters meaning dis-enclosing, bringing down the walls. An Englishman (Tom Paine again) conveyed the key of the French Bastille with its towering walls to the American George Washington. England awaits its historian of the wall, and it would truly be 'bottom up history' for was it not Bottom the Weaver who explained to Snout the Tinker how to make a chink in the wall? Well, jokes aside, the theme of livelihood is not taken up.

Can you have liberty without equality? The answer is certainly yes if liberty means liberty of property

The commons is present in the Forest Charter but the taking of commons is a story untold. Privatisation reigns quietly supreme but behind the scenes so to speak, like the whispering down the corridors of power of the English elite, the classic gentleman's agreement.

The issue seems to be this: I'm casting the exhibit as a reply to E.P. Thompson (largely on account of the title, Taking Liberties), but the issue that's really on the table is the praxis of commoning versus the privatisations of neoliberalism wherein identity politics, particularly, of women and people the colour of the earth, have attained recognition at the expense of class struggle. The wages-for-housework perspective was no less than revolutionary. Third Worldism placed such identities squarely as productive of surplus value, thus changing the notion of the proletariat to those not receiving even the 'irrational' wage. Neither new sovereign nations nor access to the franchise within the old nation liberate the proletariat any more than President Obama or Secretary of State Hillary Clinton represent victorious avatars of the working class. But what is the notion of the working class? It takes us back to 1963 and Thompson's *Making of the English Working Class*. Now as then, this notion is a political question of the future, though Thompson saw the future only through the past that he had discovered. Equality has always had a meaning in Britain as 'a man's a man for a that and a that', but the equality inherent in 'just conditions' or an equality of material access, in short, an abolition of the division between necessary value and surplus value, awaits its conclusion. Hence, the light which the Forest Charter might throw. The praxis of commoning implies fair shares of the product and fair shares in the work. The end of exploitation requires the expropriation of the expropriator.

Economic rights are present in a section entitled Freedom From Want, an American phrase. It is odd to find it here since Americans like Andy Kopkind, the

The Who and Whom of Liberty Taking

fine 1960s journalist, fled the USA to England precisely to find socialist discourse, and here is an English woman taking an American phrase to express an idea she durst not express in the idiom of its birth. The phrase was one of the Four Freedoms formulated in the context of the Battle of Britain and enunciated by Franklin Delano Roosevelt in 1941: the freedom of speech, the freedom of religion, the freedom from fear, and the freedom from want. President George W. Bush in October 2001 deliberately omitted these last two 'freedoms' in his speech on the endless war against terrorism. While Bush was preparing to do as he list, Tony Blair sat in the gallery with a pleasant look on his face, perhaps thinking of marmalade thickly spread. It's good to see the return of the phrase even as part of British liberties.

A series of *The Saturday Evening Post* essays in 1943 explained the meanings of the four freedoms. Carlos Bulosan wrote the essay on freedom from want. Born in 1913 in Pangasinan, Philippines, he came to America at the age of 16, and was kicked around from the Alaska canneries to back-bending labour in the San Joaquin valley. He spoke for 'equal opportunity to serve themselves and each other according to their needs and abilities.' 'So long as the fruit of our labour is denied us, so long will want manifest itself in a world of slaves.' His essay draws on a poem he had written three years earlier, in 1940:

> We are the men and women reading books, searching
> In the pages of history for the lost world, the key
> To the mystery of living peace, imperishable joy;
> We are the factory hands the field hands mill hands everywhere
> Moulding creating building structures, forging ahead....
>
> We are the living dream of dead men everywhere,
> The unquenchable truth that class-memories create
> To stagger the infamous world with prophecies
> Of unlimited happiness – a deathless humanity;
> We are the living and the dead men everywhere....
>
> If you want to know what we are –
>
> we are REVOLUTION!

While he changed the last word to 'marching' for the purposes of the 1943 essay, the point remains that power is not relinquished from one class to another without a struggle, nor will we common successfully in the future without a prior and

Peter Linebaugh

righteous dis-privatisation. Colley wants representative government and a written constitution. Thompson wants to prod the nerve of outrage reactivating the whole libertarian neurological memory system down the centuries. Thompson uses a physician's tendon hammer to knock your knee on the crossed leg, to see whether it is still capable of giving a kick! OK children, time to go, as the Diggers sang, 'stand up now, stand up now!'

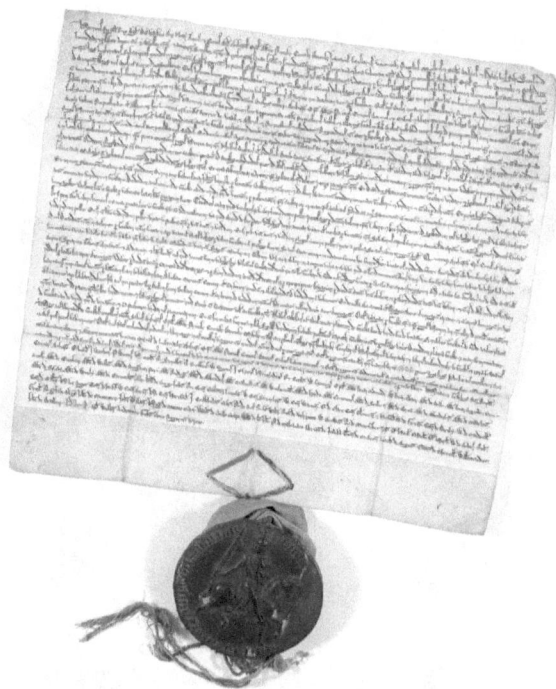

Footnotes

1 '*Corvée* is labour, often but not always unpaid, that persons in power have authority to compel their subjects to perform', Wikipedia, http://www.wikipedia.org

Info

Taking Liberties: The Struggle for Britain's Freedoms and Rights, British Library, London, 31 Oct 2008-1 Mar 2009, http://www.bl.uk/takingliberties

> **Peter Linebaugh, author of *The London Hanged: Crime and Civil Society in the Eighteenth Century* (Allen Laine 1991) and the *The Magna Carta Manifesto: Liberties and Commons for All* (University of California Press, 2008) completed his PhD at the Center for Social History at Warwick with E.P. Thompson**

Image: Laws of Forests, 11 February 1225

David Panos & Anja Kirschner's film, *Trail of the Spider*, allegorises the public-private land-grab known as 'urban regeneration' using the form of the Spaghetti Western. This is no shallow postmodern genre surfing, writes Neil Gray, but a passionate re-engagement with history for the sake of the present

DUCK! YOU REGENERATION SUCKER

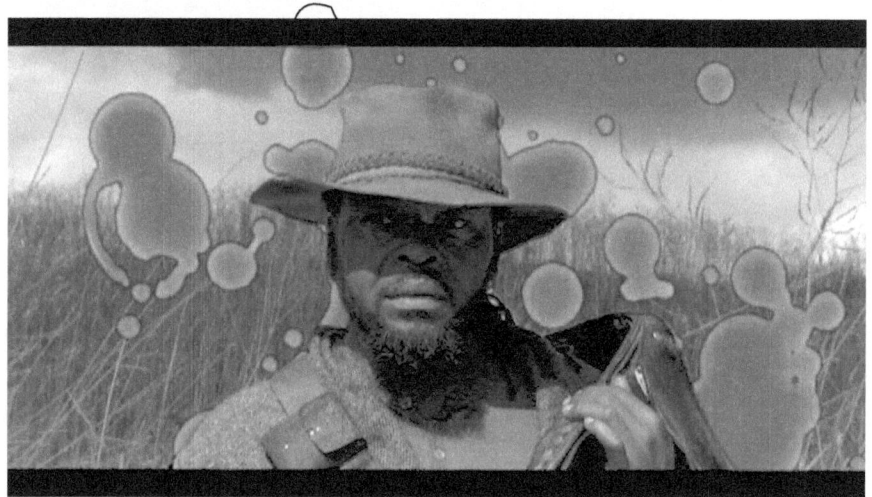

All images: stills from David Panos & Anja Kirschner's *Trail of the Spider*

> *Only that historian will have the gift of fanning the spark of hope in the past who is firmly convinced that* even the dead *will not be safe from the enemy if he wins. And this enemy has not ceased to be victorious.*
> - Walter Benjamin

> *Watch them now. All bold, like they discovered the place [...] maybe we can't stop them, but we can't give them an easy ride. Let* them *feel some loss!*
> - Doctor Dynamite, *Trail of the Spider*

Neil Gray

Stepping confidently out of the specialised ghetto of 'artists cinema', Anja Kirschner and David Panos' most recent film, *Trail of the Spider* (2008), dramatically explores the material and psychological conditions brought on by gentrification through the refractory prism of Hollywood and Spaghetti Western genres. The film, a self-described 'Western made in London and Essex', deploys the vanishing frontier motif and subversive Western genre tropes in complex yet unambiguous terms to examine contemporary 'regeneration' strategies in East London. By doing so, it skilfully mobilises multiple historical narratives into a fraught but productive relationship with the present. The universal theme of gentrification (now writ large as a central *productive* pillar of global economic strategy) re-imagined through the tropes of the Western, conjures a rich inter-textual feast for all those interested in the burdened intersection between cinema and politics.

The plot is replete with all the key motifs of the Spaghetti Western genre: an existential anti-hero; 'the Man with No Name'; a wronged woman; the inexorable forces of modernity in the form of surveyors, land barons, and the steam train; and pusillanimous townsmen, lawmen, and small businessmen seeking personal profit over collective gain. The Man with No Name wanders this unruly 'genre setting', seeking only rest and solace, while surveyors haunt the frontier lands, mapping and carving the land into new enclosures of time and space.[1] Frontier and genre conventions are deployed here not as retro gestures, or as deconstructive discourses on signifying practices. Rather, they are marshalled as part of a layered discourse dealing critically with a range of historical experiences and registers. Much of the viewing pleasure in *Trail of the Spider* is derived from this *engaged re-working* of historical and aesthetic references: an implicit challenge to an art world culture of repetitive reference and citation that almost uniformly fails to re-deploy and transform culture for radical social ends.

Stewart Home, after Guy Debord and Gil J. Wolman, has argued that the difference between *detournement*[2] and appropriation is precisely the difference between a transformational praxis and a slavish citation which refuses to theorise, never mind renovate, its historical moorings.[3] The currents which inform *Trail of the Spider*, by contrast, are rooted in the objective conditions of postwar Europe and filtered through the lens of historically situated subjectivity. The idea for the film, according to Panos, was derived in part from Munich-born Kirschner's 'very German interest in cowboys'. The devastation of Germany's production and distribution facilities during WWII granted the US film industry an industrial stranglehold on West German cinema, rendering Germany prone to a 're-education' programme in the form of re-released US films – a vast new market

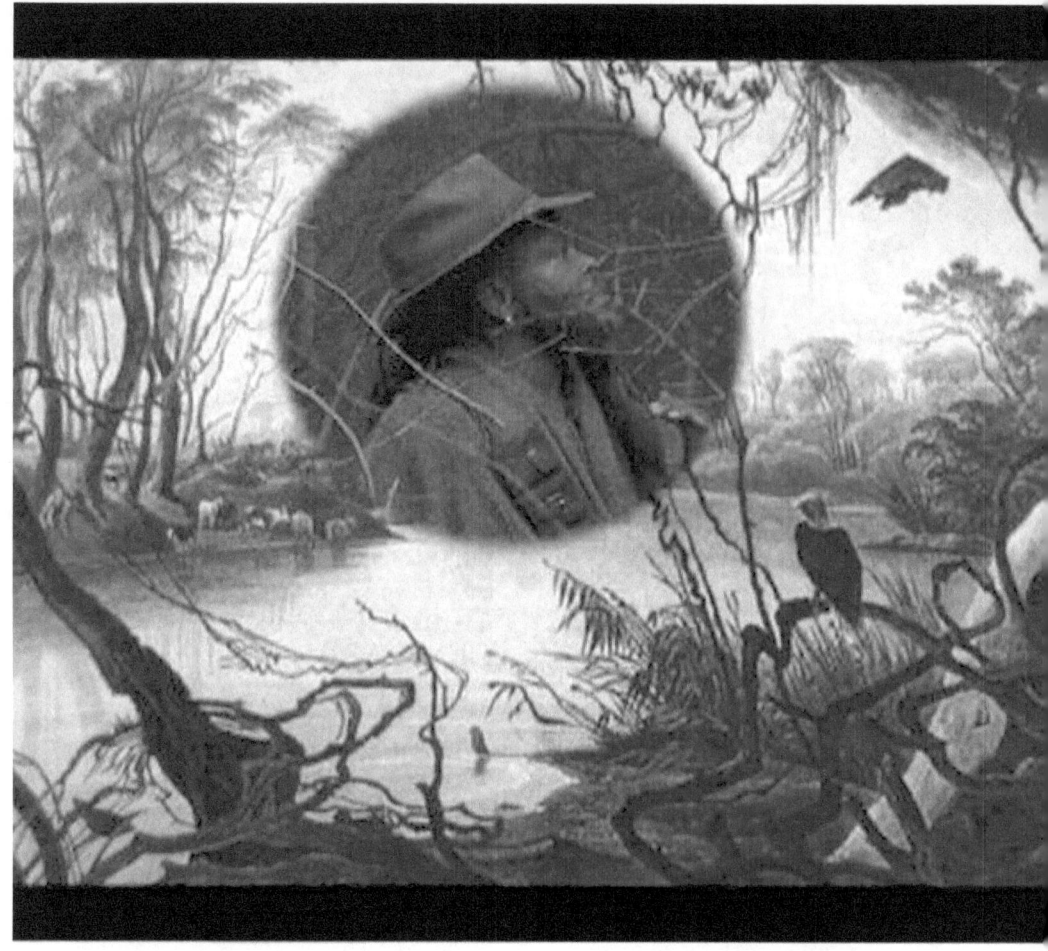

for the products of American culture. The cowboy – bolstered by the Hollywood studio system – was then the archetypal American hero; in his celluloid form, devoid of competition, he was ready to conquer German cinema. The terms of this *occupation* were disinterred by a character in Wim Wenders' *Kings of the Road* (1976) who ruefully observed, 'The Yanks have colonised our unconscious.'

The cowboy narrative, however, developed along divergent lines in the GDR, where production facilities were rebuilt relatively soon after the Nazi era. The film industry quickly developed its own ideological contours under Communist rule. Relative autonomy from US influence spawned, by the 1960s, the bastard *Indianer Filme*: an estranged, inverted breed of Western where cowboys were capitalists and Indians were proto-communists challenging North American capitalism. It is this non-compliant appropriation of the Western, shorn of regressive Stalinist ideology, that irradiates the narrative and provides a rebellious genre background for *Trail of the Spider*. The Man with No Name character (immaculately played by Hackney

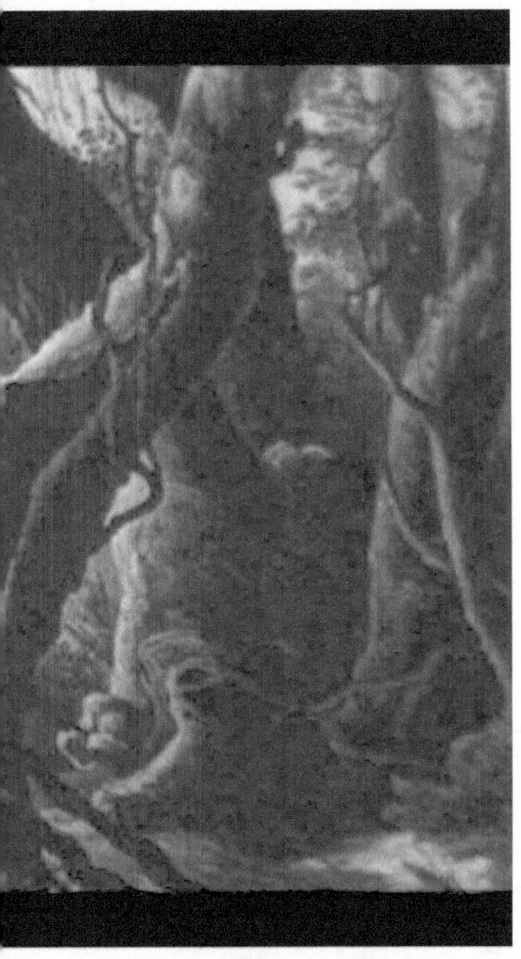

activist Floyd) finds echoes in the subversion of racial ideology in R.W. Fassbinder's *Whity* (1970), the heroic 'half-breed' avengers of Sergio Corbucci's *Navajo Joe* (1966), and Enzo G. Castellari's *Keoma* (1976). 'Marnie's place' (the tavern in Turnwood) deliberately evokes Joan Crawford's gender-busting saloon in Nicholas Ray's *Johnny Guitar* (1954), and Marlene Dietrich's 'Chuck-a-Luck' rebel hideaway in Fritz Lang's 'feminist' *Rancho Notorious* (1952). Meanwhile, references to *Django* (1966) and *The Great Silence* (1968) impregnate the film with Corbucci's 'communist' take on the Spaghetti Western genre more commonly associated with cruel, existential pathos.[4] These references (and many others) work to inscribe dissenting discourses into the prosaic conventions of genre narrative, and more importantly, into the discursive terrain of contemporary land-grab scenarios.

Yet the process by which Kirschner and Panos redeploy history and genre myth recalls most forcefully the philosophical method by which Walter Benjamin set out to rescue history from 'historicism's bordello'. Benjamin's conception of montage ('the ability to capture the infinite, sudden or subterranean connections of dissimilars as the major constitutive principle of the artistic imagination') fought a constant battle on behalf of history's victims, in order to detonate the slumbering time of the present with the fractious constellations of the past.[5] Interviewed about one of her previous films, *Polly II: Plan for a Revolution in Docklands* (2006), Kirschner said:

> To some extent the plot of *Polly II* was based on actual events from the 18th century [...] But I'm not depicting or referencing these moments so they can be measured against so many subsequent defeats or presented as easily digestible celebrations of 'heritage' or downright nostalgia [...]. Rather, I use them because they penetrate the present like so many callings and loopholes whose explosive potential still speaks to us.[6]

Duck! You Regeneration Sucker

Indianer Filme, an estranged, inverted breed of Western, provides a rebellious genre background

These 'callings and loopholes' are dramatically precipitated when the Man with No Name observes the Chief Surveyor's masked henchmen lynching a local man (the masks cite the hooded Klan gunmen in *Django*). A close-up of a wound in the Man with No Name's stomach dissolves into a rapidly edited montage that violently implodes a dense personal and collective history of repression and injustice (signified here by a lurid red filter). Later, when the Man with No Name is beaten to near unconsciousness by Turnwood for his bounty, the red filter is redeployed as a deep signifier of emotional as well as physical turmoil.[7] This defeat is given historical and cinematic resonance by the incessant, rhythmic pounding of a steam train – that potent harbinger of 19th century change – and the beating of fists, on a soundtrack pulsing with the *pressure* of brutal and seemingly irresistible forces.

Kirschner and Panos understand, just as Benjamin did, the need to fight and re-fight the battles of the past. The film re-enacts and reorders the repressed racial history of the West through scenes and inter-titles that reference a frontier history more commonly epitomised by erasure and disavowal (up to one in three cowhands were African American). The moribund landscapes of contemporary East London – through which the Man with No Name restlessly wanders – are given a mythic register and re-animated to mirror 'The Unassigned Lands' where the US government of the 1860's forced Indian tribes to 'cede back' inalienable territory so that settlers could pursue a ferocious land-grab. 'The Trail of Tears', through which the Man with No Name makes his sorrowful way to his old flame Marnie, condenses his journey with the forcible displacement and relocation of Indian tribes from the American South to Arkansas and Oklahoma after the passage of the Indian removal Act in 1830. 'The Great Dismal Swamp' (evoked in the film through judicious close-ups of snakes, lizards and spiders!) is both the location for 'Turnwood Town' and a reference to an actual location in North Carolina where runaway settlements ('maroons') of African and Native Americans escaped from slaveholders to form their own agricultural and trading communities – rebel communities which 'served as beacons to discontented plantation slaves and drove slaveholders to fuming anger.'[8]

The interaction of these mythic landscapes with present tense actuality 'brushes history against the grain', placing elements of past and present into new dialectical images that attempt to address a critical understanding of history.[9] Patterns of montage within and between images allow the landscape to be read as a palimpsest, containing multiple discursive layers. The epic Western landscape, for instance, is imaginatively conjured from landfills, 'wastelands' and gravel pits with direct links to the construction of the 2012 Olympic Park, conflating both historical and contemporary land-grabs. Meanwhile, regular insertions of 19th century paintings of the West reinforce the genre setting, and at the same time illustrate and comment on the constructed manner in which the West has always been, and continues to be, represented.

Pioneers and Pariahs – the New Urban Frontier

> If Hollywood wanted to capture the emotional center of western history, its movies would be about real estate. John Wayne would have been neither a gunfighter nor a sheriff, but a surveyor, speculator, or claims lawyer. The showdowns would appear in the land office or the courtroom: weapons would be deeds and lawsuits, not six guns.
> - Patricia Limerick Nelson[10]

The positivist idea of the frontier and its inverse relation to wilderness and 'the other' are particularly pertinent here.[11] Neil Smith charted this territory when he established the ongoing currency of Frederick Jackson Turner's prototypical 1893 essay, 'The Significance of the Frontier in American History'.[12] Turner envisioned the expansion of the Western frontier as:

> the outer edge of the wave [...] the meeting point between savagery and civilization [...] interpenetrated by lines of civilization growing ever more numerous.[13]

In this interpretation, the frontier is represented as an evocative combination of economic, geographical and historical advances by robust pioneers. Yet Turner's frontier line was extended less by individual pioneers and rugged homesteaders and more by 'banks, railways, the state, and other collective sources of capital.'[14] In the present era, the mythology of individualism persists through the *ideology* of neoliberal capitalism. By the mid-to-late 20th century, the potent imagery of wilderness and frontier was being re-applied to inner city areas in major cities back East. Inner city slums were increasingly demarcated as 'urban wilderness' or worse 'urban jungles' in a 'discourse of decline' that came to dominate representations of the inner city.[15] By the 1970s and '80s, market-led discourses of urban renaissance retooled frontier imagery (with the African American as its stigmatised 'other') to legitimise 'a political geographical strategy of economic reconquest'.[16] The property market unleashed a wave of 'urban scouts', 'urban pioneers' and 'urban homesteaders' on the margins of the inner city. And just as the original 'pioneers' reductively envisioned native Indian populations as no more than a constituent part of the wilderness, so the 'new folk heroes of the urban frontier' saw the residents of the contemporary frontier as 'not yet socially inhabited' or 'less than social'.

Duck! You Regeneration Sucker

Trail of the Spider exhumes this discursive current of conquest and legitimation from the opening scene. In a terrific panning shot across a massive, moribund East London landfill site, a voiceover, culled from one of the Communist *Indianer Filme*, works with and against the image to stridently conflate the ideas of frontier conquest, American capitalism and ruthless displacement with current speculative land-grabs in the East End of London. The mythologised tracts of contemporary London stand in for the historic Unassigned Lands (crackling with effect-laden lightning) where the remnants of a splintered resistance eke out a hounded existence in a rapidly diminishing social space; a space where surveyors, surrounded by 'security', chart 'the hitherto unmapped territory' of potential capital investment. Gazing out over the land for investment opportunities, the Chief Surveyor (in an assuredly pompous performance by Robin Laine), propounds the real meaning behind the reductive 'civilization/wilderness' binary:

> Look at the lay of this great land. Who would have thought that one day we would attach value to this wilderness?

The film-makers have stated that the main concern of *Trail of the Spider* was to allegorise and question the 'the shifting and shrinking space for collective social and political agency, self-determination and dissent' in an urban reality increasingly dominated by large-scale urban gentrification – what Neil Smith has ominously, if somewhat a-historically, described as an 'unassailable capital accumulation strategy for competing urban economies'.[17] Like Benjamin and his close friend Bertolt Brecht, these concerns are expressed with an understandable degree of historical pessimism (Benjamin and Brecht's thought in the '30s, after all, was shaped by the degeneration of the USSR and the seemingly inexorable victories of fascism). Yet those unpropitious times also grounded an immanent, dialectical view of history, divested, for the most part, of ideological myopia. As Brecht once recommended: 'Don't start from the good old days, but the bad new ones.'[18]

Kirschner and Panos understand the need to fight and re-fight battles of the past

In these *bad new days*, the relevance of *Trail of the Spider* should be readily apparent. 'Sugar-coated' promises of community 'regeneration', can't hide the fact that what's really occurring is gentrification on an unprecedented scale (with all it's ugly connotations of class displacement). The prosaic reality of rampant property speculation is dramatised near the conclusion of the film, when a contemporary, fictionalised 'land race' on an open tract of land in London's East End is compared

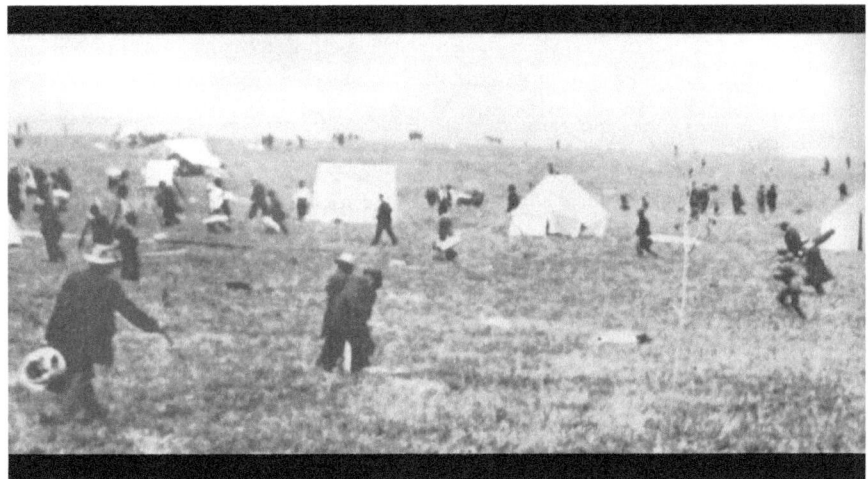

(through the insertion of archival photographs) with the Oklahoma land race of 1889. The event saw Indian squatters forcibly removed from land that had been promised to them so that white settlers could 'compete' for plots of land. The conflation of these narratives of speculation and enclosure, with all the legalistic and judicial apparatus that profits from them, nakedly reveals the consistent current of brutal, competitive self-interest that lies behind capitalist accumulation strategies. Displacement, advanced marginality, and the ruthless disposal of public assets are the necessary contingencies of this totality.[19]

Hardt and Negri usefully theorise representative democracy as a 'disjunctive synthesis' that simultaneously 'connects and separates' citizens from the social body.[20] In this context resistance is typically either forced to the margins or defused through legalistic, reformist channels. In both *Polly II* and *Trail of the Spider*, Kirschner and Panos acknowledge and question the problem of recuperation and mediation. But rather than merely contemplating their own alienation, their creative practice emerges from praxis in everyday struggles against gentrification in East London. Most of the people in the film are friends or people they've met through activist campaigns in Hackney – including the Broadway Market occupation.[21] During the occupation, they met Floyd (The Man with No Name), and a prior friendship was cemented with John Barker ('Doctor Dynamite'), whose experience with the Angry Brigade in the early '70s adds a critical, self-reflexive dimension to an intimately staged campfire scene (the dialogue concerns the potential of sabotage in the face of repression and retreat).[22] Meanwhile, Marnie's impassioned speech (singer/performer Claudette Bonney in a great performance) is partly based on Spirit, a local Jamaican shopkeeper and neighbourhood friend facing eviction orders in a rapidly gentrifying Hackney. This social dimension of *Trail of the Spider* concurs with Brecht and Benjamin who saw *friendliness* as a 'minimum programme of humanity'.**23** At the same time, it locates their film practice alongside Peter Watkins, one of Britain's most marginalised and vilified film-makers. Watkins work with non-professional actors has consistently attempted to break down the petrified, alienating separation of artist and audience by encouraging the public to directly participate in representations of history – 'past, present and future'.**24**

David Panos described a recent packed-out screening of *Trail of the Spider* in Hackney which 'somehow recreated the reception to the Spaghetti Westerns in Jamaica and other post colonial countries'. Such a reaction echoes the scene in Perry Henzell's *The Harder They Come* (1972), where the fugitive anti-hero (Jimmy Cliff) watches Corbucci's anti-hero *Django* (a major influence on *Trail of the Spider*) annihilate a gang of racist henchmen with an enormous sub-machine gun. The wild amusement of the crowd is undoubtedly borne from the Jamaican audiences recalcitrant experience of independence struggles against British colonialism. *Trail of the Spider* seeks a similar kind of resonance with its audience. In an aesthetic register that confronts the ossified conservatism of high modernism, and the mindless quotation of postmodernism, the film also brushes popular genre conventions against the grain in an attempt to engage and re-arrange the audience's experience and knowledge of history. This way of working with history isn't mere source material for a reified and institutionalised art practice; instead historical forces are represented in such a way that they provide a genuine motivational connection to the possibilities of social change. As Nietzsche said in *The Use and Abuse of History*, 'We need history, but not the way a spoiled loafer in the garden of knowledge needs it.'**25** If Nietzsche's sentiment is true for history, it is also true for a contemporary urban reality in desperate need of radical transformation.

Info

Trailer for *Trail of the Spider* (2008):
http://www.anjakirschner.com/trailtrailer.html
Trailer for *POLLY II - Plan for a Revolution in Docklands* (2006):
http://www.difficultfun.org/items/pollytrailer.html

Footnotes

1 The phrase belongs to Paul Willemens. Paul Willemens, 'An Avant-Garde for the '90s' in *Looks and Frictions: Essays in Cultural Studies and Film Theory*, British Film Institute, Indiana University Press, 1994, p.157.

2 Defined as 'The reuse of preexisting elements in a new ensemble' in 'Detournement as Negation and Prelude', p.55, Ken Knabb ed., *Situationist International Anthology*, the Bureau of Public Secrets, 1981.

3 Stewart Home, *Bubonic Plagiarism: Stewart Home on Art, Politics & Appropriation*, Sabotage Editions, p.10.

4 Akira Kurosowa's denuded Marxism led him to the pessimism of *Yojimba* (1961), the chief inspiration for Sergio Leone and the Spaghetti Western genre. No wonder that Kurosowa's exotic 'depth', shorn of his earlier politics, appealed so much to Hollywood luminaries George Lucas and Steve Spielberg.

5 From Stanley Mitchell's penetrating introduction to *Walter Benjamin: Understanding Brecht*, Verso, 1998, p.xiii.

6 http://www.anjakirschner.com/polly2/reviews.htm

7 A device that vividly evokes the cinema of Michael Powell and Emeric B. Pressburger, for instance, *Black Narcissus* (1947) and *Peeping Tom* (1960), and their interest in the irrational frictions lurking beneath the carapace of the rational mind.

8 http://www.anjakirschner.com/trailofthespider/trailpamphlet.pdf

9 Walter Benjamin, *Illuminations*, Pimlico, p.248.

10 Patricia Nelson Limerick, *The Legacy of Conquest: The Unbroken Past of the American West*, New York: Norton, 1987, p.55.

11 For an analysis of how frontier imagery is mobilised as a neoliberal alibi for creative destruction of inner city areas: in this example, the East End of Glasgow. See Neil Gray, *The Clyde Gateway: A New Urban Frontier*, Variant, Issue 33, Winter 2008, (forthcoming online).

12 Neil Smith, '*The New Urban Frontier: Gentrification and the Revanchist City*', Routledge, 1996.

13 Ibid, p.xiii.

14 Ibid, p.xvi.

15 The phrase is borrowed from Robert Beauregard, *Voices of Decline: The Postwar Fate of US Cities*, Oxford: Basil Blackwell, 1993.

16 Neil Smith, op. cit., p.xvii.

17 Neil Smith, 'New Globalism, New Urbanism' in *Spaces of Neoliberalism: Urban Restructuring in North America and Western Europe*, Blackwell Publishing, 2002, p.96.

18 Walter Benjamin paraphrasing Brecht's critique of Lukács on realism. 'Conversations with Brecht' in *Walter Benjamin: Understanding Brecht*, Verso, 1998, p.121.

19 Advance marginality is described, somewhat long-windedly by Loic Wacquant, the originator of the term, as: 'the novel regime of sociospatial relegation and exclusionary closure [...] that has crystallised in the post-Fordist city as result of the uneven development of the capitalist economies and the recoiling of welfare states [...]'. Loic Wacquant, *Urban Outcasts: A Comparative Sociology of Advanced Marginality*, Polity Press, 2008, p.2.

20 Michael Hardt and Antonio Negri, *Multitude*, Penguin Books, p.231-247.

21 For background see, http://www.metamute.org/en/The-Re-Occupation

22 For a highly readable account of this period see Stuart Christie, *Granny Made Me an Anarchist: General Franco, The Angry Brigade and Me*, Scribner, 2004, p.318-411. See also John Barker, *Transgressions: A Journal of Urban Exploration*, Issue 4, Spring 1998, Salamander Press, p.101-107, for Barker's own take on the legacy of the Angry Brigade's actions (online: http://www.geocities.com/pract_history/barker.html).

23 Walter Benjamin on Brecht's poem, 'Legend of the Origin of the Book Tao Te Ching on Lao Tzu's Way into Exile' in *Walter Benjamin:Understanding Brecht*, Verso, 1998, p.74.

24 See, http://www.mnsi.net/~pwatkins/part2_home.htm for more on Watkins method and filmography.

25 Quoted in Walter Benjamin, *Illuminations*, Pimlico, p.251.

Neil Gray <neilgray00@hotmail.com> is a writer and film-maker based in Glasgow

> Giving a critical survey of the documentaries of Adam Curtis, Andrew Fisher evaluates the claims to realism and political neutrality made for his work, using the critical methodologies of Guy Debord and Georg Lukács

THE SLEEP OF REALISM PRODUCES MONSTERS

My job is not to try to change the world, but to describe it.[1]

Working at the BBC since the late 1980s, Adam Curtis has become one of the most celebrated contemporary British documentary film-makers. He is routinely fêted as the author of ambitious films that offer self-consciously provocative viewpoints on contemporary social and political issues. Since 2001, he has made three television series – *The Century of the Self* (2002), *The Power of Nightmares* (2004) and *The Trap: What Happened to Our Dream of Freedom* (2007) – each of which outlines a lengthy and generalised account of ideas, individuals or elites understood to have had a formative impact on the present.[2] Together, these films attempt to understand the present, but what understanding they offer remains politically ambiguous and thus demands critical analysis, not least because it is riven by a tension between the films' emphatic claim to realism and their play on the spectacular form of mass media.

Curtis views history as a 'series of unintended consequences resulting from confused actions', in which context things never work out as intended.[3] The idea of historical narrative he derives from this is defined in opposition to the notion of 'balance' that characterises the institutional framing of journalism (especially at his

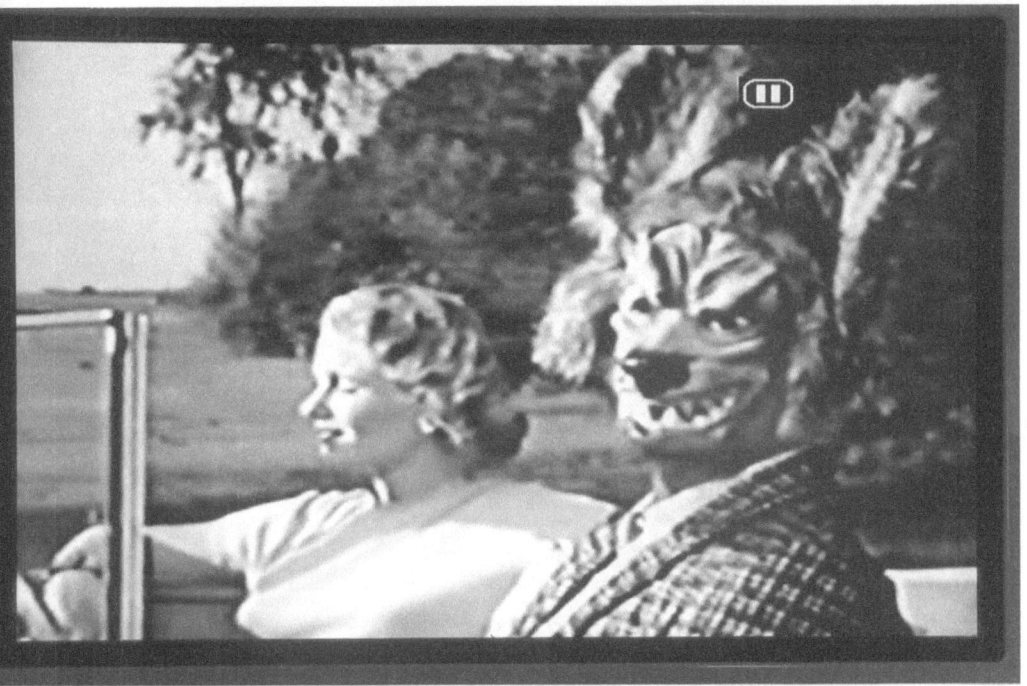

Image: Adam Curtis, *The Century of the Self*, 2002

place of work, the BBC), namely the injunction that one must show 'both sides' of a story.[4] This idea of balanced representation tends, in Curtis' view, to frame events in a formal symmetry that all too easily prefigures and restricts the bases upon which critical judgements about them might be made. His response to this conventionalised mediation of history is to knock it off balance, to narrate it polemically and in slanted terms.

The Century of the Self charts the influence of Freud's discovery that irrational forces are at work in the human psyche and seeks to demonstrate how, as Curtis frames it, 'those in power used Freud's theories to try to control the dangerous crowd in an era of mass democracy.' The series gives a reductive account of Freudian psychoanalysis, as imported to America by Freud's nephew Edward Bernays, marketing entrepreneur and apparent inventor of public relations. An often entertaining narrative of the rise and fall of the influence of psychoanalysis in America ends with an account of its collapse into irrationally determined individualism in the '60s, and the rest of the series charts the apparently more 'scientific' (and objectifying) measures of behaviour that emerged from corporate attempts to turn the newly minted individuals of the post '60s period into newly active consumers. A key figure of this relation

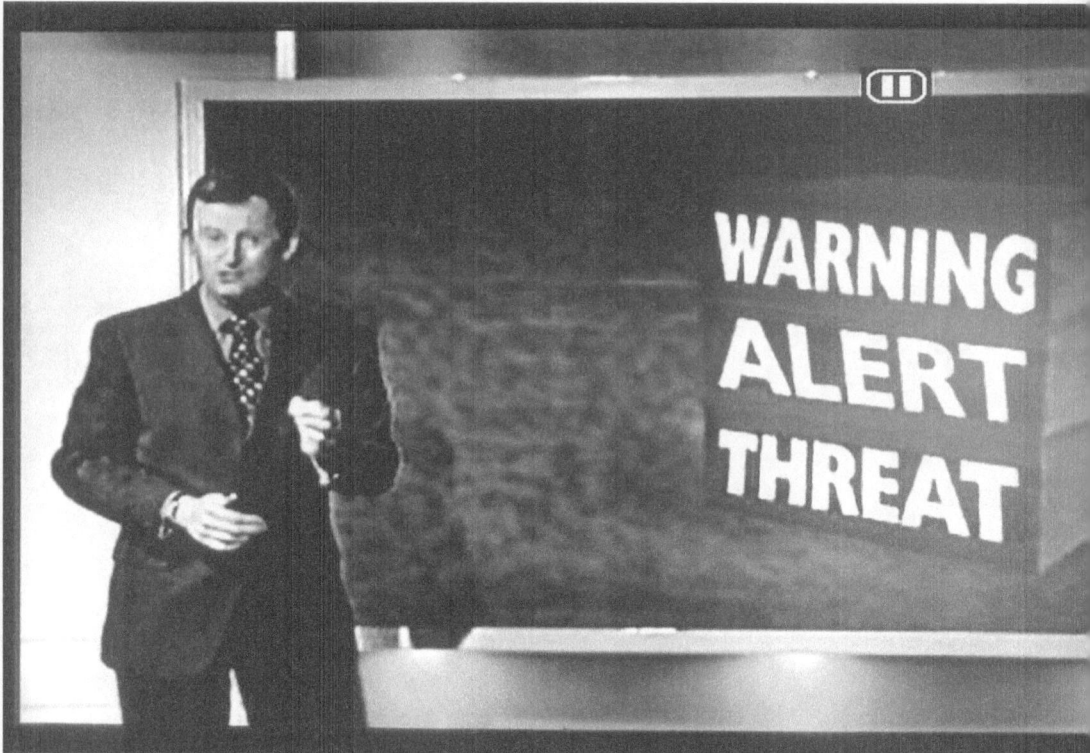

Image: Adam Curtis, *The Power of Nightmares*, 2004

is Werner Erhard – author of 'Erhard Seminars Training' (EST) – the promoter of a nigh-on total rejection of the social in favour of a selfish fantasy about the 'redemptive' personal possibilities of radical self-interest. It is with a critically framed account of the intimate relation between the apparent opposites of cynical self-interest and objectifying behavioural measurement that the series concludes.

In *The Power of Nightmares* (the most interesting of these series), Curtis traces ideas that have informed neoconservatism and radical Islamism in order to speculate on their co-dependence. The intellectual roots of these movements are found in a shared distaste for western liberalism that he reads out of the writings of the Chicago philosopher Leo Strauss and the Egyptian revolutionary Sayyid Qutb. For Curtis, these two embody broad trajectories inspired by the sense that liberal individualism threatens moral decrepitude and promises social chaos. Out of this linkage emerges a tale of how American and British governments and politically revolutionary Islam ended up on opposing sides of a world-encompassing imaginary sphere, polarised by a cynical and shared commitment to the myth of an 'evil other' that defines one's own values as good and one's acts as right. This quite familiar way of thinking about America, in particular, is given a controversial twist as Curtis takes his narrative about the fantasies of fear that characterise the War on Terror to call into question some of its central figures, such as the actual existence of al-Qaida, Osama bin Laden's role as its leader, and the reality of the danger presented by 'dirty bombs'. His claim is that these evils have been made up, constituted in exaggerated form through the uneasy and ambiguous mediation of actuality as a media spectacle oriented to produce fear and to sustain it in the interests of social control.

These documentaries are best understood in terms of the tension they foreground between a realism that seeks to excavate the significance of individuals (e.g. Bernays) and ideas (e.g. the myth of the evil other), and the strategic play on the spectacular form of the media through which these facts are presented. Whilst it is not, I suspect, a description with which Curtis would readily concur, his films take up, mix and concentrate Lukácsian themes (the power of literary realism and the prevalence of reification) and the influence of the Situationist International (reification accelerated in the spectacular form of the commodity image and the possibility this holds out for its critical *détournement*).

In light of this, it's worth dwelling on the formal character of his narratives. Each film is structured by a text written and spoken by Curtis over montaged sequences of images drawn from various archival sources. By a number of means, these foreground the fact that what one is seeing has already been subject to mediation (such as in his repetition of a questionable cinematic image of an Arab magician in *The Power of Nightmares*). Montaged archival material is contrasted to interviews conducted by Curtis and filmed using available light and limited sound equipment in a way that makes what the interviewees say seem unmediated. All of this is set to often ironically counterpoised music (as in one sequence

Curtis' films do not share Debord's understanding of the political imperatives of détournement

from *The Trap* in which the title 'F**k you Buddy' is screened over a clip of a singing, stars-and-stripes bedecked Bing Crosby). These forms of layered juxtaposition and different stylistic approaches to more or less obvious forms of mediation stress a self-conscious display of the fragmentary and polysemic character of narrative constructions. However, the range of possible associations projected by particular juxtapositions tends to be closed off by the neatness and the directive character of his voiceovers. In this sense, Curtis' practice bears comparison to Guy Debord's film *The Society of the Spectacle* and the similar way it frames archive images and music with a polemical narrative. For Debord, however, such culturally reflexive strategies ought to aim at a targeted political critique. This is what, in his view, gives polemical appropriation, montage and repetition meaning as strategic modes of opposition to the spectacular form of the commodified image. For Debord, this holds out the promise of a cultural practice that is something more than a self-conscious inflection of the commodity form. Crucially, in a society for which 'all that was once directly lived has become mere representation', the *détournement* (appropriation and re-functioning) of conventionalised culture has to risk 'speaking the language' of spectacle in order to render its dubious certainties uncertain and its apparent necessities contingent.[5] Whilst Curtis' films are, perhaps indirectly, indebted to *The Society of the Spectacle*, they clearly do not share in Debord's understanding of the political imperatives informing such strategies. Curtis trades, sometimes skilfully, on the sense in which the situationist idea of the spectacle has tended to be politically denatured by those who embrace it as a description of contemporary culture.

> **Curtis aspires to a complex narrative that Lukács identified with the realist novel**

The polemical confidence of Curtis' narratives (there's no question mark in the title of *The Trap: What Happened to Our Dream of Freedom*) serves to establish his films' attempt at realism. He frequently relates them to 19th century literary realism, as in the following:

> You look at Bleak House, say, and Dickens throws 10 strands of news stories together and sees where they lead him. I love that.[6]

It's not going too far to say that he seems to aspire to the kind of complex narrative totality that Lukács identified as the determining characteristic of the realist novel.[7] But Curtis imposes a rather crude Lukácsian framework on the historical processes he charts. In *The Century of the Self*, for instance, he exploits a range of archival sources to good effect such as an old car advert in which a man imagines himself as a sexualised wolf driving a new convertible, demonstrating acutely just how cynical and thinly disguised advertisers' attempts to manipulate desire have been. However, despite its incisive moments, this history of conflicts over the idea of the self condenses them into the biographical figure of Bernays, the genius of manipulation, who – by lucky application of a traduced Freudianism to commercial culture – comes to typify the relation between self, society and history in an arbitrary and jaundiced manner. Curtis simplifies the complex relationships in question. He boils down the processes charted into this figure, articulated in a tightly knit narrative whole. This echoes Lukács' understanding of literary characterisation and the imaginative, temporally complex manner in which it might typify a social

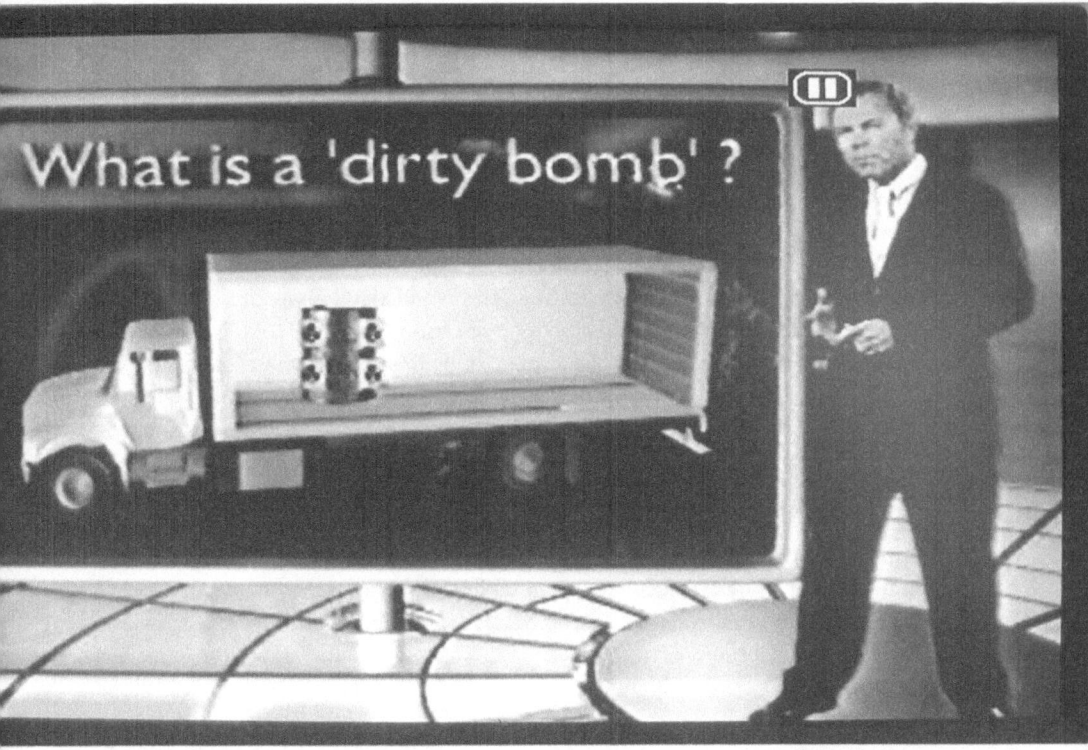

Image: Adam Curtis, *The Power of Nightmares*, 2004

subject struggling to make sense of their historical context.[8] For Lukács, the dialectical relationship between character and the formation of narrative totality introduces fruitful temporal complexities into the extended present of readerly experience. In Curtis' realism there is no such dialectic and the temporal character of the present appears static. Whilst his Lukácsian framework serves to delineate some striking characters and to speculate upon the influence they might have had on the present contemporary society, it also, quite heavy handedly, suggests that the idea of a present which is open to change is a thing of the past.

Thus, in a sense, the structure of Curtis' films separates out the functions of narrative and montage in their constitution of emphatic narrative totalities centring on singular ideas, individuals and elites. While looking for a way to understand his treatment of history and the form of Curtis' realism, one might think of these relations between text and image, montage and narrative as performing different operations: the archive and the spectacular character of the image produce complexity whilst the narrative text attempts to guarantee the sense of the whole and its relation to the facts. But thinking of this as an imbalanced realism, which might be quite promising, also begs consideration of the problematic things Curtis does with it.

The Trap: What Happened to Our Dream of Freedom is perhaps the most ambitious and, arguably, the most problematic of Curtis' post-2001 works. Conceptually and formally it builds upon the earlier series and makes more emphatic their tendency to crude generalisation. Starting from a consideration of Freidrich Hayek's political philosophy, *The Trap* attempts to establish an intimate duality. Curtis explores, on the one hand, the RAND Corporation and John Nash's development of the distinctly paranoid paradigm of Game Theory – here, an embrace of the Cold War logic of Mutually Assured Destruction extended to describe human behaviour – and, on the other, the claim that this desperate logic found social confirmation in R. D. Laing's adoption of a similar framework to describe the family as the destructive site for the production of mental illness. A set of wide ranging (sometimes dubious) conceptual connections allows Curtis to chart the application of management theories, based on Game Theory's core notion of selfishness, to most areas of public life. He frames this in terms of Isaiah Berlin's opposition between positive and negative notions of freedom; the former being a seductive utopian projection of ideals that are taken to lead inevitably to terror and repression, and the latter, defined in opposition to this, as a limited notion of freedom expressed in the minimal organisation of society to facilitate indulgence of personal desires.

In this framework, the anti-institutional celebration of individualism in '60s and '70s culture is described as dovetailing neatly with the formal abstractions and modes of self-interest characteristic of forms of social administration inspired by Game Theory.[9] That, in this process, politics has become management means, for Curtis, that negative freedom has been realised in dystopian form. The film ends with a plaintive call for a political imaginary that strives to be positive in Berlin's sense but without producing repression. This plea is contradictory as the preceding episodes accept at face value Berlin's assertion of the impossibility of realising this desire. Insofar as it is a reprise of the central themes and conclusions outlined in the earlier series, *The Trap* reveals all three films to be melancholically structured around a notion of the dissolution of political agency, but they make little attempt to describe what the conditions of (or the precedents for) such agency might be, that is, not until near the end of each series where the value of politics and subjectivity have apparently been dissolved. Ultimately, *The Trap* is structured by a pessimism that concedes the 20th century to those who want the world to have realised itself according to Berlin's anti-speculative paradigm.

Curtis' works have provoked considerable debate, ranging from their rejection as conspiracy theory by conservative commentators to left wing celebrations of their criticism of neoconservatism.[10] What links and motivates these responses is the idea that his films are political, but this understandable assumption is not as straightforward as it might seem. For instance, *The Power of Nightmares'* reception by the left needs to be questioned given that Curtis' reconstruction of the pre-history of the War on Terror offers, in the end, most sympathy to Henry Kissinger's brutal political pragmatism.¹¹ Curtis has repeatedly distanced himself from the idea that his aims are political, as in the following contrast he made in 2004: '[Michael Moore's] purpose is avowedly political. My hope is that you won't be able to tell what my politics are.'¹² Albeit quite casually, this contrast echoes a familiar and intransigent dilemma facing left culture: whether to aim at a better conception of the past or a more forcefully desired future as the structuring principle of one's intellectual labours. Whilst Curtis' films do resonate with many of the political left's concerns (for instance, privatisation and neoliberalism, networked organisation, capitalism's obsession with risk and security, identity politics and the authoritarian personality), the interest of his engagement with these themes is undercut by the claim that his project isn't political. But this claim cannot simply separate politics from the meaning of Curtis' films. Rather, it gives a clue to their diffuse political character. For one thing, it entails accepting the idea of the separation of political agency from the institutions and processes the films narrate and ostensibly oppose, which is disingenuous. For another, it disappoints the more critically interesting moments his approach has produced.

There's a passage in *The Power of Nightmares* that relates the story of a widely reported 'terror alert' in 2002 that resulted from the CIA's and FBI's 'debriefing' of Abu Zubaida (apparent operations chief of al-Qaeda) and his vague descriptions of plans to attack New York that were inspired by the 1998 remake of the film, *Godzilla*. Curtis' treatment of this odd story is interesting. He narrates it as an instance of the co-dependent and mythical projections of an 'evil other' that are, significantly, mediated in the globally disseminated American remake of this film (an already heavily coded Japanese fantasy of nuclear destruction, in which previous myths, and realities, of evil and otherness lie sedimented). Against this story are set a sequence of images: broadcast news reconstructions of anti-terror police in action against pictures of slack jawed television audiences; a Guantanamo inmate against the dated special effects of an old Aladdin movie; Godzilla causing havoc on the streets of New York against a White House spokesman relating the news of Zubaida's threat to an alarmed press corps. Those who ridicule this film for its claim that the key figures in the War on Terror are made up (and those who take it to be straightforwardly realistic) might look again at this sequence to note the way it builds on the exaggeration, duplicity and cynicism of such layering of myth upon myth and the way that this amounts to what one might call a *speculative* image that adopts such a form as a way of making concrete the truth of an

> while Curtis' films resonate with the left, he claims that they aren't political

actuality that is highly ambiguous. Such moments are occasionally achieved in Curtis' films. Unfortunately, they have to be critically excavated from a framework that seems all too happy to accept his religious and conservative interviewees' assertions that speculative (utopian and idealist) thought is doomed. Curtis' films are caught in a liberal dilemma which views speculation as a positive ('big ideas' form the core of all his films) but believes this value to be obsolete in a world taken to be correctly described by those whose interests are served by the erasure of an ability to imagine things otherwise. There are different qualitative moments of speculation in these films. Sometimes these lie in what Curtis says (e.g., his plea for a renewed politics in *The Trap*) and sometimes in his construction of complex images (as in the *Godzilla* sequence). I think the pessimism of the former is best thought separately from the promise of the latter.

Footnotes

1 Adam Curtis interviewed by Andrew Orlowski for *The Register*, 20 November 2007, at, http://www.theregister.co.uk/2007/11/20/adam_curtis_interview/page6.html

2 After earning his journalistic stripes (presumably sourcing talking dogs) for Esther Rantzen's *That's Life!* in the '80s, Curtis went on to produce a number of documentary series for the BBC, including *Pandora's Box* (1992) and *The Living Dead* (1995). It is in the account in *The Mayfair Set* (1999) of a small elite of 'buccaneer' businessmen and their influence on the development of Thatcherism that his characteristic approach to documentary narrative took shape and this has been consolidated in subsequent series.

3 See 'It becomes a self-fulfilling thing: Adam Curtis talks with Errol Morris' at, http://errolmorris.com/content/interview/beleiver0406.html

4 For instance, see his comments in conversation with Morris: 'I'm very suspicious of this idea of a balanced version of history. All history is construction – often by the powerful. What I do is construct an imaginative interpretation of history to make people look again at what they know. [...] Because I think that so much of this interpretation of events is a deadening repetition agreed upon by certain people, a sort of collectivity of news reports.' Ibid.

5 Guy Debord, *The Society of the Spectacle* (1967), trans. Donald Nicholson-Smith, New York: Zone Books, 1995, Thesis 1, p.12.

6 Quoted by Tim Adams in 'The Exorcist', *Observer*, Sunday October 24, 2004 at, http://observer.guardian.co.uk

7 See, George Lukács, 'Realism in the Balance', in Ernst Bloch, et al., *Aesthetics and Politics*, ed. Ronald Taylor, London: New Left Books, 1977, pp.28-67.

8 See, for instance, George Lukács, 'The Intellectual Physiognomy in Characterisation', in *Writer and Critic and Other Essays*, ed. & trans. Arthur Kahn, London: Merlin Press, 1978, pp.149–88.

9 This formulation draws closely on Brian Holmes' discussion of the films in question here. See, 'Neolib Goes Neocon: Adam Curtis, or Cultural Critique in the 21st Century' at, http://www.nettime.org/Lists-Archives/nettime-1-0706/msg00047.html

10 A sense of the range of critical responses to Curtis' work can be gained by looking at the following selection: 'The Making of the Terror Myth', Andy

Andrew Fisher

Beckett, *The Guardian*, Friday, 15 October 2004, at, http://www.guardian.co.uk/media/2004/oct/15/broadcasting.bbc; Clive Davis, 'The Power of Bad Television: The BBC's Bizarre New Documentary on Terrorism and Neoconservatism', *National Review*, 21 October 2004, at, http://nationalreview.com/comment/davis200410211043.asp; 'Beware the Holy War: The Power of Nightmares', Peter Bergen, *The Nation*, 20 June 2005, at, http://www.thenation.com/doc/20050620/bergen; Kenneth Minogue on *The Trap* at, http://www.socialaffairsunit.org.uk/blog/archives/001440.php; 'Cry Freedom', Oliver Burkeman, *The Guardian*, Saturday, 3 March 2007, at, http://www.guardian.co.uk/media/2007/mar/03/broadcasting; 'The Trap', Anindya Bhattacharyya, 10 March 2007, Socialist Worker Online, issue 2041 at http://www.socialistworker.co.uk/article.php?article_id=10878; David Black's review of *The Trap*, 10 June 2007, at, http://www.thehobgoblin.co.uk/journal/h92007_db_trap.html

11 See Curtis' comments to this effect in the conversation with Morris referenced above.

12 Quoted by Andy Beckett in *The Making of the Terror Myth*, op. cit.

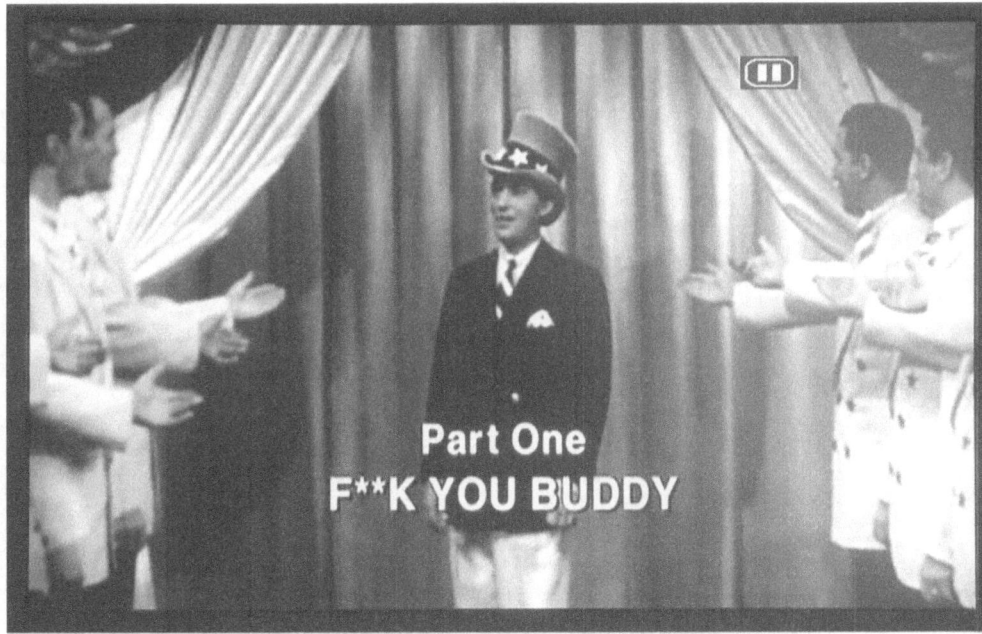

Image: Adam Curtis, *The Trap: What Happened to Our Dream of Freedom*, 2007

Andrew Fisher <hss02af@gold.ac.uk> is an artist and writer who lectures in the Visual Cultures Department of Goldsmiths College, University of London

Subscription offer

10% off the Mute catalogue

Subscribe to Mute and guarantee to be the first in line for our quarterly collection of provocative articles on culture, politics and technology. What's even better, subscribe now and not only get Mute delivered straight to your door but receive 10% off of our new Catalogue, including magazine back issues and titles from the OpenMute print on demand press. That's a year or more of discounts on books, back issues and Mute special projects. Below is just a small selection from the broad range of products on offer.

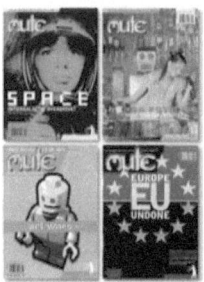

Find yourself a Mute short of a full set? Take advantage of our subscriber offer to get 10% off. If you missed out on earlier formats, Mute Back Issues collections offer sets of back issues for only **£35**/collection (which means **£31.50** if you subscribe). Mute's new collections – grouped according to the magazine's successive formats – make it easy to build or complete your very own Mute library*.
- Back Issues I: the Broadsheet (pilot-issue 7, safely packed in two unique pink folders, 'back pack 1 & 2')
- Back Issues II: the Glossies (issues 8-24)
- Back Issues III: Coffee Table (issues 25-29)
- Back Issues IV: the POD (current volume, issues 0-7).

*We regret to say that none of these include issue 9, which is now sold out.

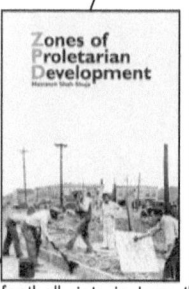

Shah-Shuja Mastaneh's *Zones of Proletarian Development*
Zones of Proletarian Development is an attempt to theorise the anti-capitalist movement from a neo-Vygotskian perspective. It analyses a series of proletarian activities including recent May Day celebrations in London, carnivalesque football riots in Iran, the anti-poll-tax rebellion and the anti-war movement. Concluding by looking at past and current proletarian organisations, this book makes a number of proposals for future modes of organising conducive to radical consciousness and autonomous activity.
Price £15 **£13.50**

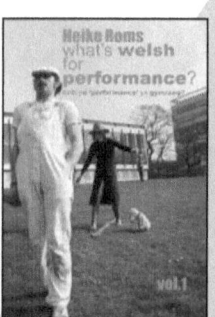

Heike Roms' *What's Welsh for Performance? Beth yw 'performance' yn Gymraeg?* **An Oral History of Performance Art in Wales 1968-2008**
For more than forty years artists have been creating performances, happenings and other time-based art in Wales, yet their work remains largely confined to half-remembered anecdotes, rumours and hearsay. *What's Welsh for Performance?* tries to uncover Wales's hidden history of performance in conversations with key artists who have shaped this history since 1968.
Price £10 **£9**

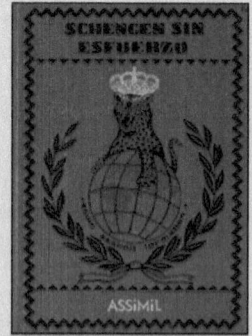

Vahida Ramujkic's *Schengen with Ease*
'Extra-comunitarios', or citizens of non-European countries, have the 'extra' bureaucratic task of changing their status to one that will allow them to move and work 'freely' within the European Union. All the required steps are taught through lessons like those found in foreign language skill books, comparing the administrative language of European immigration legislation to an unknown language that has to be mastered first in order to assimilate in to a new environment, receiving determined status.
Price £8.29 **£7.46**

Download the complete catalogue at metamute.org/catalogue
Or contact lois@metamute.org +44 (0)20 7377 6949 for a printed copy

metamute.org/catalogue

mute

Subscription Rates:

	individual		institutional/company	
	4 issues (1 year)	8 issues (2 years)	4 issues (1 year)	8 issues (2 years)
uk	☐ £20	☐ £38	☐ £35	☐ £67
europe	☐ €28	☐ €52	☐ €48	☐ €91
usa/mx	☐ $40	☐ $75	☐ $70	☐ $133
row	☐ €34	☐ €60	☐ €54	☐ €102

Please tick the appropriate box.

I wish to pay by cheque/credit card.

☐ I enclose a cheque (GBP) made payable to Mute.
☐ Please charge my

☐ Visa   ☐ Access   ☐ Mastercard   ☐ Switch

Card no. ☐☐☐☐ ☐☐☐☐ ☐☐☐☐ ☐☐☐☐

Expiry date ☐☐ / ☐☐

[Switch only] Issue number ☐☐   Start date ☐☐ / ☐☐

Security code ☐☐☐

Signature _____

name _____
address _____
_____
_____
town/city _____
post code _____
country _____
tel _____
email _____

Or call our credit card hotline
Tel +44 (0)20 7377 6949
Fax +44 (0)20 7377 9520

Online metamute.org/shop
Email mute@metamute.org
Skype mute.london

POST TO: MUTE,
Unit 9, The
Whitechapel Centre,
85 Myrdle St.,
London E1 1HQ, UK

www.ingramcontent.com/pod-product-compliance
Lightning Source LLC
Chambersburg PA
CBHW020445220526
45464CB00002B/858